The
Figure
Finaglers

The Figure Finaglers

ROBERT S. REICHARD

Senior Economics Editor, Purchasing World
Member of Business Research Advisory Committee,
Bureau of Labor Statistics
Member of Faculty, New York Institute of Technology

McGRAW-HILL BOOK COMPANY
New York   St. Louis   San Francisco   Düsseldorf   Johannesburg
Kuala Lumpur   London   Mexico   Montreal   New Delhi
Panama   Paris   São Paulo   Singapore
Sydney   Tokyo   Toronto

Library of Congress Cataloging in Publication Data

Reichard, Robert S
  The figure finaglers.

  1. Commercial statistics. 2. Statistics.
3. Consumer education. I. Title.
HF1017.R43   381'.3   73-21643
ISBN 0-07-051777-0

  234567890 BABA 7654

*The editors for this book were W. Hodson Mogan, Robert
E. Curtis, and Gretlyn K. Blau, the designer was
Naomi Auerbach, and its production was supervised
by George E. Oechsner. It was set in Alphatype
Vladimir by University Graphics, Inc.*

Printed and bound by George Banta Company, Inc.

To Haze, Ali, Pete, Dora, Etta and Teddy

This is the age of the consumer. After decades of neglect, the American public is finally coming into its own—aided in large part by a spate of legislation in such consumer-oriented spheres as safety, packaging, advertising, and ecology. Viewed in proper perspective, these recent advances are all part and parcel of what for want of a better phrase can be termed the "consumer revolution."

But up until now one key area—statistical reporting—has been neglected. For more years than we care to admit, the public has been hoodwinked by number-spouting "con men" and charlatans whose main purpose in life seems to be pulling the wool over some unsuspecting layman's eyes. Indeed, the sheer number of quantitative distortions now seems to have reached the epidemic stage, threatening to (1) mislead the innocent and (2) instill an intense dislike for anything statistical or quantitative on the part of the more informed.

Both are dangerous trends. Correcting the former, of course, should be an economic as well as moral must. But the growing tendency on the part of many to suspect anything statistical could be an even more serious problem over the longer run. For, make no mistake about it, constructive quantitative analysis is needed. Indeed, it is virtually indispensable for the running of what is fast becoming a number-oriented, automated society.

The problem, then, is not to junk statistics, but rather to insure its constructive use by consumers—to separate fact from fiction, the useful from the useless and misleading. It is to this task that *The Figure Finaglers* addresses itself.

In a sense the book treads on new ground, for a study of the existing literature suggests that little work has been done in this critical area. To be sure, there are several frivolous books on how to lie with numbers. But they fall short on several counts. For one, they're meant more to amuse or entertain than to inform. More significantly, such works make few attempts at adequate subject coverage. Three or four obvious distortions are blown up, and that's the sum total of all the really useful information contained in such books.

The Figure Finaglers on the other hand, represents the first attempt to rigorously classify, document, and analyze the literally hundreds of ways an unsuspecting user of statistics can be duped. Emphasis is on (1) asking the right questions, (2) recognizing the telltale signs of duplicity and distortion, and (3) making the public aware of the limitations and pitfalls of any quantitative approach.

While the subject matter concerns itself with numbers, the book is basically nonmathematical in approach. Formulas and the like are used only sparingly, and then only when necessary to point up how figures can be used to distort or mislead. As such, anyone with a good high school education and reasonable intelligence should be able to handle the subject matter with ease.

The approach and writing style also are geared to the non-technical reader. Thus the book is faithful to the language and flavor of the everyday lay world, emphasizing what is important to the consumer rather than what interests the statistician or mathematician in his ivory tower. Moreover, to inject the crucial note of credibility, the examples used have been culled from the real world. Indeed, in almost all cases the reader will be able to recall similar situations from his own experience.

The hope is that when the reader has completed the book he will have at least a nodding acquaintance with every weapon in the statistical charlatan's arsenal, and will thus be better prepared to meet the onslaught in any one of the many fields where statistics are now being used to deceive.

In short, *The Figure Finaglers* isn't aimed primarily at entertaining, but rather at injecting some common-sense rules into what heretofore could only be described as a statistical jungle. Hopefully, the ideas contained will provide the impetus for another great leap forward in the meaningful march toward full consumer sovereignty.

Robert S. Reichard

ONE

HACKING IT THROUGH THE STATISTICAL JUNGLE

Figures don't lie, liars figure.

Chances are you've heard this old bromide before—it's been making the rounds for years now. But whether you have or not, it still has the ring of truth, for there's obviously nothing inherently wrong or dishonest in the use of numbers by itself. Rather, the problem lies in how these numbers are used to deceive, manipulate, and distort.

Lest there be any doubt on this score, just open up your favorite newspaper or tune in your local television channel—or, better yet, think a moment about that socioeconomic argument you were having with your best friend the other day.

More likely than not, there were innumerable examples in each

1

of the above where statistics were used to prove a point rather than to provide an unbiased picture of the real world.

The newspaper columnist writing about inflation will undoubtedly choose those types of prices that show a sharp advance—and conveniently forget to mention any areas where prices are stable or even declining. Similarly, the hard-sell TV commercial will quote all sorts of irrelevant statistics to convince you that brand A is far and away the best brand on the market.

And even on the sensitive and controversial subject of racial integration there are those, on both sides of the issue, who will readily produce whole sets of numbers designed to prove their respective, self-serving points. No doubt there are those who will have statistics ready to prove how, say, busing has helped the socially disadvantaged, whereas there are others who will have an equally impressive set of numbers to show that this controversial approach has actually cut down the rate of learning and increased racial polarization. Neither of these approaches presents an entirely accurate picture of the facts.

Politics, too, offers ample opportunities for playing the numbers game. Consider the familiar use of numbers attending a political rally to establish the popularity of a given candidate. If candidate A attracts 20,000 people, then candidate B shoots for 40,000. And if a pro-war demonstration attracts 10,000 cheering partisans, the antiwar forces try for 15,000 or even 20,000.

To a large extent, this amounts to little more than statistical distortion. It's pretty easy to manufacture crowds. Just schedule a motorcade at lunchtime through a crowded business district— or let the kids out of school—or even bus in supporters from outlying districts. In short, a large crowd more often than not may be a sign of the skill and energy of the organization arranging the rally rather than a manifestation of the broad spontaneous support it purports to represent.

Finally, even the actual numbers being used should be suspect. How do you measure the mass of humanity crowded into a town square? Depending on who's doing the counting, the numbers can vary significantly. Indeed, when you get right down to it, most such crowd estimates come right out of thin air.

All of the above illustrations have one thing in common: the use of quantified data to prop up an argument rather than to establish the ultimate truth. In a sense, playing the numbers game is part of human nature. It's the statistical apple in the proverbial Garden of Eden. Why not take a bite? No one's looking. No one's going to get hurt. Indeed, it's the rare individual who hasn't been tempted at one time or another to take some data out of context in order to establish a crucial point.

And yet numbers are indispensable in today's modern technologically oriented world. There can be no doubt, for example, that current progress, in business as well as in the physical and social sciences, is in large part due to the collection, tabulation, processing, and analysis of quantitative data. Numbers are the lingua franca of the modern world. In short, they're here to stay.

The problem, then, is not to abolish the quantitative approach, but rather to establish the rules of the game, to assure that figures are used to enlighten rather than to cloud the crucial and not-so-crucial issues of our day. A way must be found to make people aware of how they're being duped—and, more important, how they can turn the tables on their oppressors and actually use numbers and figures to upgrade their decision-making processes.

It is to these goals that this book addresses itself.

THE NUMBERS MYSTIQUE

Before going into some of the more detailed do's and don'ts, it might be well to dispel some of the myths and distortions about numbers.

Actually, there are two schools of prevailing thought. One involves the true believers—a large group who accept almost any quantification of data as gospel. Their philosophy: If it can be expressed as a number, it has to be correct.

Then there's the opposite school: those who tend to believe that there's a lie behind almost every statistic. This popular barb of the nonbelievers perhaps best sums up their philosophy: People use statistics like a lamppost—more for support than illumination.

Of course, neither of these two extreme approaches to numbers and figures is correct. Take the true-believer school first —those who stand in rapturous awe of each new bit of quantified information that comes across their desks. This is part and parcel of our tendency to accept anything we see in print as the gospel truth, simply because it has been printed. Let the government issue a preliminary figure on retail sales or prices, and some people are ready to accept it as revealed wisdom. Then, too, when the power of massed rows and columns of numbers is buttressed by the force of the computer, criticism passes the irreverence stage and borders on rank heresy. This attitude is by no means rare—especially in the executive suite. Having spent several million dollars for a new computer, many businessmen are loath to admit that it can do any wrong.

Part of this, of course, is due to lack of sophistication about what the computer can and cannot do.

Computers can't do the thinking for you, nor can they create something out of nothing. More specifically, it should be immediately apparent (but many times isn't) that the GIGO—or garbage-in, garbage-out—approach to quantitative analysis is equally applicable to counting on one's fingers and using an expensive new computer.

Without good input, good output is impossible, no matter how people may try to rationalize otherwise.

This input type of problem crops up when social statistics

are involved. All too often there's a tendency to quantify what is essentially an unquantifiable concept. Result: The numbers may be accurate enough, but they are essentially meaningless.

Indeed, the validity of many statistical series which were accepted without question a decade ago is being increasingly questioned. In almost all cases they involve the equating of statistical fact with absolute truth. This can best be explained through a concrete illustration.

It may be possible, for example, to ascertain that student A did better than student B on an intelligence test. But to assume that this means student A is smarter than student B is to assume the infallibility of intelligence tests—something which in most cases seems to be unwarranted.

In a similar vein, it is a fact that during the Vietnam war the United States had an extremely favorable "kill" ratio relative to that of the North Vietnamese. But the simple assumption that this meant we were winning the war in Vietnam was quickly disproved.

It all boils down to the fallacy of equating shadow with substance—and the surprising number of distortions and self-deceptions that can result from this approach. (Chapter 5 will treat this important area in more detail.)

A few further words are also in order on the nonbelievers—those who summarily dismiss the quantitative approach as meaningless or a hoax. These people, in addition to pointing out the problem of quantifying social concepts, also like to magnify some of the shortcomings of the statistical approach. But they conveniently forget all the impressive pluses yielded by detailed analysis of numbers and figures.

The *average* is usually the favorite whipping boy of these critics. One of their typical snide remarks is: "The average American shaves half his face and powders the other half."

This kind of remark, often born out of unhappy experience, plays up the ludicrous. It ignores the great deal of useful information that such a simple yardstick as the average can provide. It also gives short shrift to the fact that averages are the key building blocks for many of our sophisticated statistical approaches.

The use of anecdotes to emphasize the misuse of numbers is still another favorite ploy of the statistically disenchanted. Hence the popularity of the story involving a two-car auto race between the Russians and the Americans. The Americans win. But the report in Pravda reads as follows: "The Russian car finished second, the American car, next to last."

Then, too, these doubters like to point up the fact that numbers are never as precise as they're made out to be. This is, of course, true. But few reputable statisticians would ever make such a claim. One hundred percent accuracy would be neither feasible nor practical, for perfection in statistics, as in almost any other type of endeavor, is an impossible goal.

FAKING THE FACTS

On the other hand, the more vociferous critics of the statistical approach do have a point in cases where numbers are used to create the illusion of quantification in areas where such quantification is physically impossible to achieve.

The crowd count example referred to on page 2 is a case in point. But there are many other such bogus numbers flung around with wild abandon by people who either have an ax to grind or want to establish an aura of expertise in an area where no such expertise is possible.

Thus how many times have you heard someone talk about the number of people going to bed hungry at night, the number of rodents in a city slum, the billions of GNP dollars lost through a flu epidemic or a hurricane, the aggregate cost to the nation of traffic jams—or even the dollar value handled by organized crime syndicates?

Consider the problem of measuring the rodent population. A

few years back one group of investigators in a large Eastern city actually tried to count the number of rats in a small section of a slum—and then extrapolated this figure for the entire city (a highly questionable technique at best). Result: They figured there was one rat for every 40 or so people living in the city. Another study, this one taken by a government agency, however, came up with a much different figure—one rat per every 5 people.

How did the "powers that be" resolve this eightfold difference in estimates? They came up with the seemingly profound—but really meaningless—statement that the true rat density probably lay somewhere in the range between one rat for 5 and one rat for 40.

First, the range is so wide as to make the figures virtually useless. But more important, one could raise the question, How does one enumerate the rodent population? The investigator may walk through the tenement counting the number of rats he sees on each floor. But this is a highly questionable procedure at best. How does he know he has seen each rodent on each floor? How many have been hiding in the woodwork or the plumbing? Or how many, for that matter, has he counted twice? Rodents certainly don't go around with identification bracelets. Then, too, rats, unlike people, are mobile. It may be the same little fellow you've seen on the first, second, and third floors as he follows you on your census-taking rounds. The point of all this, of course, is that there is absolutely no way of knowing how many rats there are on a floor or in a building—and especially in any given area of a city.

Similar estimates about stray dogs or stray cats appear in papers at least once every few months. Again the question may be asked: It's hard enough to take a census of people; how does one take a census of dogs or cats, who will hardly stand still long enough to be counted? Again the answer is that it simply can't be done.

Leaving the animal scene for a moment, consider the problem of estimating the cost of traffic jams. If the loss is estimated at $15 billion a year, how does one define "loss"? Is it the aggregate value of output lost? Is it sales lost because of late delivery? Or what? But beyond the definitional question, how could anyone really know how many traffic jams there are in a given year, how many vehicles are involved, and how long the jams lasted?

One so-called expert, when asked to explain his estimate, came up with the lame answer, "It seems to be in line with past experience." That's double-talking at its best.

Losses involved in storms are equally questionable. The problem here is twofold. First, nobody seems to take the trouble to define what they mean by "loss." It could be the original cost of the equipment destroyed. Then again, it could be the replacement value — particularly if it is the one who suffered the loss who is asked to estimate the damage. On the other hand, if the insurance company does the estimating, it will more than likely be the depreciated value that is used. (Anyone who has had his car stolen can attest to the insurance company philosophy on just what constitutes fair value.)

An equally difficult problem arises in estimating the actual physical volume of damage or loss. How does one go about enumerating all the losses? Insurance claims are often cited as the basis for such estimates. But how does one build up from insurance claims to all losses — for surely, in a storm as large as a hurricane, uninsured losses can be 10, 50, or even 100 times as large as the insured losses? But who knows what the true multiplier really is?

Then there are the estimates of gambling and crime losses — also equally suspect. The investigating committee that claims the syndicate take is, say, $1 billion a year from the numbers game in New York — or that the take of all organized crime in the United States is, say, $50 billion — is probably doing little more than guessing.

One thing for sure, it's hard to see how they can come up with any meaningful figures, inasmuch as organized crime has never been known to report their earnings to the Internal Revenue Service.

The list could go on and on. But perhaps the most blatant form of the phony count approach involves the journalist or politician who uses these figures. He usually gets them two or three years after the supposed count was taken. So he feels justified in beefing them up by 10 percent or 20 percent to take into account subsequent growth and inflation. Moreover, he may not be satisfied with the "impact" effect of the figure—so he may well add on another 25 percent or so, saying that "informed sources" have indicated that the original estimate was on the low side. Nowhere, however, does he identify these "informed sources."

In short, watch out for the phony, the misleading, and the unsubstantiated figure. Whenever the fiction writer or journalist needs statistical documentation to justify a story, be assured he'll come up with some expert who can provide it; for if he doesn't, he won't have his story, his expertise, or, for that matter, his job.

OTHER CAVEATS, ERRORS, AND DECEPTIONS

Problems are compounded by the fact that there are literally thousands of ways of pulling the wool over the unsuspecting layman's eyes. The phony statistic approach noted above is only one of many. In a sense, the highways and byways of deception are limited only by the imagination of the perpetrator.

Semantic differences, for example, can often be used to cloud over the real meaning of numbers. The current argument over desegregation provides a perfect illustration. As will be shown in Chapter 10, it can be established that we have achieved desegregation on a purely de jure or legal basis. But clearly we still have a long way to go if a de facto definition of the word is used.

During the Vietnam war we were plagued with the problem of semantics. Thus in the summer of 1972 newspaper reports headlined the fact that the troop strength had dropped under the 50,000 mark.

Strictly speaking, this was true. But what these same releases failed to note was the fact that while the Vietnamese troop withdrawal was going on, a buildup was taking place among United States navy and air force personnel fighting the Vietnam war from bases in Guam and Thailand and on ships off the Vietnamese coast. It took a lot of reading between the lines to discover that this "other" Vietnamese force had at that time been built up to over 100,000 men. In short, the government was taking advantage of a narrow definition to give the impression that we were withdrawing from Southeast Asia.

The above is not to imply a value judgment on whether such troops were needed in Southeast Asia at that time. Rather, the illustration is meant only to show how statistics did give many people a misleading impression of what was actually happening.

But semantic differences hardly explain all problems. Sometimes distortions of data exist but are unintentional—as the result, in large part, of the user's lack of sophistication. Take the big 20 percent jump in 1972 Social Security benefits. Most oldsters were quick to jump to the conclusion that their overall benefits would increase the full 20 percent. But this was not always the case.

In some states, assistance to Social Security pensioners was reduced by the amount of the increase. In other places the federal increase made some pensioners ineligible for state medicaid programs or reduced their food stamp allotment. Ergo, many oldsters ended up with considerably less than a 20 percent boost.

Another popular misconception of the 1970–1972 period: the belief that individuals had gained more than businessmen from federal tax cuts. This belief was fostered by the quoting of income tax rates which showed the sharpest cuts on the personal income tax level.

But what was left out was the fact that the effective federal rate on corporate profits had fallen significantly because of (1) an investment tax credit and (2) liberalized depreciation allowances. Add these two in, and the effective rate on corporations dropped from 41 percent to 37 percent in the 1970–1972 period. By contrast the 1969 and 1971 tax relief for individuals didn't fall nearly as sharply.

Another example of unintentional distortion. The interpretation of monthly cost-of-living changes. Because of widespread reporting in the press, the government price measure has become a household word. The trouble is, however, that these reports are analyzed in many cases by people who don't have the slightest statistical qualification to interpret the results.

Indeed, the way the price trend is reported in the press could lead to the conclusion that it is fantastically erratic. Let the cost of living rise 1 percent, and some are led to believe that runaway inflation is upon us. Let prices decline the next month, and these same interpreters would have us believe that the nationwide inflation problem has finally been licked.

Political pressures compound some of these distortions. Every slight price decline is heralded by government officials as a great victory. But let the same index creep up a bit, and these same officials will call it a "temporary statistical aberration"—and that one month's rise is hardly meaningful.

Sometimes the government will even change its price calculating ground rules to bolster its position. Thus in 1972, Uncle Sam's price watchers reversed themselves over the treatment of antismog devices. Originally, they had ruled that the cost of putting these devices on automobiles should be reflected, cent for cent, in government price indexes. But this tended to raise price indexes at a time when Uncle Sam was waging an all-out fight to bring them down. Ergo, a flip-flop. Government statisticians, after due deliberation, decided that the new device was more of a "quality improvement"—something that doesn't show up in official indexes as a higher price (see Chapter 12).

There's another more insidious deception practiced by government. The built-in bureaucracy becomes so enamored of the numbers it is turning out that it loses sight of the larger goal—the use of such numbers for meaningful policy-making decisions.

In a sense it is self-deception which is being practiced here. And it has perhaps been best put by Arthur M. Ross, a former Commissioner of Labor Statistics. Said Ross a few years ago: "Statistics consists of technical procedures quite independent of content or purpose. I found that most government statisticians are principally concerned with techniques, which have greatly improved in recent decades. But their outlook is often too narrow to encompass the larger role of numbers in public life."

Put another way, many of the problems today stem from a lack of understanding about what numbers can and cannot do. The inclination in too many cases is to take what can be quantified or measured, not ask any questions, and assume that such statistics represent objective truth.

Thus anything under a given income is equated with poverty; a given "kill" count is equated with success or failure in waging a war; and a given GNP level is equated with happiness. It is only over the past few years that we have begun to realize how mistaken we can be by taking this easy, uncritical way out.

Sometimes just sloppy press reporting is at fault. Thus in 1972 a committee of the American Medical Association charged that the automobile accident statistics of the kind headlined in the press on holiday weekends are misleading and possibly inaccurate. It singled out, particularly, the claim that drunken drivers cause more than half of all fatal crashes. This should be changed to state "that 50 percent of all motor vehicle crashes in which there are fatalities involve alcohol," the committee claimed.

"For instance," the group explained, "a number of studies have shown that about 10 percent of these fatal crashes involve a drinking pedestrian whose demise very likely was his own fault and not that of a drinking driver." In some instances, a drinking driver may have been involved in a fatal crash that was caused by a nondrinking driver, it added.

Another myth commonly accepted by the public: the belief that the gap between the rich and poor has been narrowing. Nowhere do figures back this thesis up.

Indeed, data from a recent study show that in almost every year since 1947, the poorest fifth of American families has received only about 5 percent of the country's total family income, while the top fifth got 42 percent—an 8:1 ratio.

In still other cases, propaganda can make people choose against their self-interests. Another recent survey (this one conducted by the Advisory Commission on Intergovernmental Relations) indicated that 46 percent of the people favored a sales tax as the best way for states to raise more money. Only 2 percent favored the more progressive income tax, which would have left a majority of respondents with more real purchasing power.

At still other times, mere laziness or the pressures of time or money lie behind such distortions. We suspect something is wrong, but there just isn't enough time or funds to search out the correct figures. So we deceive ourselves in saying that what's available is an approximation of what "true" figures would show.

Change, either sudden or gradual, is still another source of statistical misunderstanding. Too often there's a tendency to assume that what held yesterday will hold through today and probably into tomorrow as well.

Here both the businessman and the consumer fall into the same trap. A copper wire producer who estimates his own firm's current demand by looking at past wire consumption would in all probability overestimate his sales by three or four times. Reason: recent technological developments which have permitted the substitution of cheaper aluminum for copper in the big wire market.

At other times change is unpredictable. Fashion, for example, can play hob with projected numbers. Take, for example, the horrendous flop of the long (midi) skirt in 1970. Manufacturers, following the suggestion of fashion designers, put all their money on the acceptance of the midi. But their projections that year proved to be way off, as the consumer stuck with the old miniskirt. Result: widespread manufacturer losses, with many outfits forced into bankruptcy.

Why the midi length failed in 1970 came out in subsequent interviews with women. Generally speaking, the consensus at that time was that the long skirt made fat girls look fatter, tall girls look dumpy, and older women look older. One woman even likened the midi to the Edsel—a Ford automobile that flopped miserably during the early 1960s.

There are many other ways in which numbers can deceive, too. Percentages, the old popular standby of advertisers, politicians, businessmen, and even educators, lend themselves to a whole series of distortions—some subtle and others not so subtle. Interest rates were, of course, up to recently the most blatant example of the misuse of percentages. A few years ago, few consumer borrowers could tell you the true rate of interest they were being asked to assume when taking out a loan.

The variations on the percentage theme are such that two full chapters will be devoted to this family of problems.

Another popular route to deception: the old "apples and oranges" routine. Unless comparable phenomena are compared, the results are likely to be meaningless. Retail sales in summer can't be realistically compared with sales in the pre-Christmas December season. Comparing market with list prices can lead to equally serious distortions. Then too, profits of company A may well be under those of company B—not because A is run any worse, but only because each uses a different depreciation method.

Spuriousness is still another type of pitfall to look out for when dealing with quantitative data. What does "average family income calculated to the nearest penny" really mean? What does the advertiser mean when he says his product is "20 percent better" than his competitor's?

Use of numbers to establish cause and effect can also lead to spuriousness. The fact that people with high-cholesterol blood content suffer a higher incidence of heart attacks doesn't necessarily mean cholesterol is the cause. It could be that the factors that lead to high cholesterol also lead to heart attacks. A similar argument can be made about cigarette smoking and cancer.

Numbers expressed in the form of charts and pictures can also be used to deceive and distort. If the numbers are wrong, then the chart depicting them is obviously wrong. But there's an additional deception that's possible in this area. By juggling the size and shape of the chart, an irrelevant point can be magnified and the unpalatable made to seem more palatable.

The above are but a few of the problems to be discussed in the following chapters. But all have one thing in common. Given the facts, all such distortions become relatively easy to spot. And, once spotted, it's possible to eradicate them with surprisingly little effort.

Summing up, the consumer need not be saddled with the indecencies of deceit, deception, and outright lies. But to do something positive, he needs (1) government help, (2) more knowhow, and, last but not least, (3) the will to fight back.

THE CONSUMER CAN FIGHT BACK

All these factors must be present. If there's any doubt on this score, look at recent experience with consumer interest rates. Without enacting legislation—forcing bankers to spell out the true rate being charged—little could have been done to redress the lender-borrower balance.

But this alone would not have been enough. Without the accompanying education and publicity it is doubtful whether the average family would have noticed the difference—or have been aware that it was entitled by law to know the true or effective interest rate.

Lenders didn't help matters too much, either. Many of them obeyed the letter of the law by giving the effective rate—but ignored the spirit of the law by burying the pertinent information far inside the contract, in "small print" and in places where borrowers were not likely to look.

But even educating the public would not have been enough. To make the law work, consumers would have to want to make it work. Indeed, even today the "truth in lending" results are mixed. Middle-income and other informed groups concerned about their welfare are now generally paying lower rates than before the legislation was enacted.

But results in low-income, "apathetic" areas have been somewhat disappointing. These people are still fair game for usurers —and are still paying far more than they have to for the privilege of borrowing money. The basic problems here and what is being done to alleviate them are spelled out in Chapter 4.

Consumer organizations are still another avenue for fighting back. Ralph Nader, the consumer advocate, may have exaggerated in many cases, but probably for the first time in their life he made buyers realize that they were entitled to protection.

Indeed, the pendulum has swung so far the other way that many claim the old "caveat emptor" (let the buyer beware) concept is dead, and that it has been replaced by a new "seller beware" approach.

In a sense this was long overdue, for as recently as five years ago a buyer could go into a store under the impression that the purchase of, say, an economy size of a product would save him money. But a little simple calculation would more often than not prove him wrong.

The author on one occasion was tempted to buy a 10-ounce

jar of instant coffee at $1.64 under the misleading impression (engendered by advertising) that by buying this size he would be a lot better off than by buying a 6-ounce jar at 99 cents. Only after calculating the cost-per-ounce price did he realize that there was very little difference.

Specifically, the 10-ounce jar yielded a unit price of 16.4 cents per ounce—virtually the same as the 16.5 cents-per-ounce quote on the smaller size. It has been only over the past few years —with new unit pricing laws—that the true costs have been exposed.

Another favorite manufacturer and supermarket ploy is to promote individually wrapped convenience sizes for "a few cents more." But the "few cents more" phrase is a woeful understatement, because in most cases the buyer is probably getting a lot less volume for his money.

Take the case of breakfast cereals. The unsuspecting buyer who opts for the "few cents more," convenient, individually-wrapped packages may be paying more than twice as much per ounce for the cereals purchased.

To be sure, the new unit pricing ground rules now in effect in many communities hasn't made this kind of two-price system illegal. But it has made the cents-per-ounce price known to the consumer who wants to know. So, again, motivation becomes almost as important as the mandatory aspects of the law.

The sharpness of the counterattacks on such "truth" laws, meantime, suggests that, regardless of arguments to the contrary, the vested interests of the vendors are indeed being affected. Both lenders and supermarkets, for example, have fought the new moves on the grounds that they would (1) only serve to confuse the consumer more and (2) create fantastic amounts of paper work.

These arguments are, of course, suspect. Specifically, why now were these vendors concerned about "confusing" a buyer whom they had been trying to confuse for years? Equally significant, their concern over "paper work" has never really stood the test of time. The additional cost of spelling out interest rates, for example, can be measured in terms of pennies.

Supermarket people were guilty of still another kind of distortion a few years back. In March 1972 these vendors agreed to freeze meat and other food prices after consumers had complained about a 6 percent (at annual rate) rise in the cost of living that month.

The press, as might be expected, greeted the move as "statesmanship of the highest order." But were the supermarkets giving anything away? Not really. For wholesale food prices had fallen that month, and these retailers could very well predict that the chances of any further rise on the consumer level were just about nil.

There are other avenues for consumer justice, too. Some of the ones that have been bearing fruit are: (1) court suits, (2) refusal to pay a bill, and (3) complaints to local Better Business Bureaus. Up till recently none of these were particularly effective—because of our seller-oriented society.

Not so today. More and more these approaches are bringing results—primarily because of changing attitudes. If you can show how a shopkeeper distorted price or other terms of sales, chances are you can get some redress.

Then there are the government agencies designed to protect the consumer against statistical deception. Only now are they beginning to do the job they were originally set up to do. These agencies include the Federal Trade Commission, the Pure Food and Drug Administration, and the consumer affairs bureaus mushrooming in almost all important population centers.

In recent years these and other agencies have been hacking away at deceptive selling practices—particularly, unfair advertising. Thus claims of the type suggesting that brand A aspirin is twice as effective as brand B are no longer permitted. Nor can an advertiser promise a buyer a given percentage of savings unless, of course, he can back up the statement with dollars-and-cents documentation.

Utility regulatory agencies also have begun to crack down. Take a recent phone company example. For more than five years millions of phone company customers had been ignoring a chance to save a sizable amount of money on special services such as melodious chime bells (instead of the old-fashioned ring) and the so-called "Princess" phone.

Reason: The phone company, in making these options available, had tended to play down the fact that customers had the option of (1) a flat fee when the service was originally installed or (2) a regular monthly service charge. Almost all advertising pitches stressed only the latter. In New York, the company pursued this policy until the state Public Service Commission put its foot down. The Commission ruled that the first option had to be given equal emphasis. When this was done, a lot of customers opted for the flat fee—often realizing considerable savings.

Take the option of the Princess phone. In 1972 a customer could pay $43.03 for a Princess phone, when installed, or pay $1.17 a month. If he paid the monthly fee for 37 months, the total cost would be equal to the flat fee. But on a monthly basis the consumer would keep paying indefinitely; under the flat-fee offer he would pay only the $43.03 and, in fact, would be able to take the phone with him when he moved without extra charge.

Thus anyone expecting to keep the Princess for more than three years would clearly be wise to pay the lump sum. Similarly, other special services would cost nothing after a certain number of months—assuming, of course, the buyer opted for the lump sum approach. Thus a "Trimline" phone would pay for itself in 52 months, volume control in 37 months, chimes in 46 months, and tone control in 52 months.

To sum up, these and other kinds of statistical legerdemain—

the kinds that have been yielding extra dividends to sellers for decades—finally may be on the way out. The following chapters, by putting the spotlight on the literally thousands of still common statistical deceptions, can possibly bring this day of final victory a little bit closer.

TWO

THE CHARLATANS OF CHARTING

One picture is worth a thousand words. There're few who would quarrel with this venerable observation. But just as a picture (or graph or chart) can be many times more effective in imparting information, so can it be many more times more effective in distorting or twisting the truth.

No wonder, then, that anyone bent on deceiving the consumer will rely so heavily on bogus or misleading charts. The most obvious example, of course, involves the direct plotting of distorted data onto a chart. Indeed, this compounds the felony, because the visual approach tends to play up any misleading information.

But for those with an ax to grind, it is possible to distort even if the underlying figures are reliable and accurate. All the statistical miscreant need do is to present the data in such a way as to help the reader reach the desired (and often misleading) conclusion. It is the intention of this chapter to focus on the second type of misrepresentation—the kind that can be traced back to the choice of chart rather than to any inadequacies in the basic data. Some of the more notorious examples of graphic distortion are discussed below.

THE FLOATING OR "NON-ZERO" BASE

This is a devilishly simple technique whereby the chartist conveniently forgets to put his numbers in proper perspective. He zooms into (or magnifies) the area he wants to emphasize, leaving other relevant information completely out of the picture.

Fortunately, there's always telltale sign of this kind of foul play: the absence of a "zero" line on the chart or graph. Thus, if you want to impress a wage earner on how well off he is with a series of four $500 annual pay boosts (say, from an initial base of $10,000 per year to $12,000 per year), plot $10,000 at the bottom of the chart and $12,000 at the top of the chart. Your income then seemingly goes from the depths (see bottom of Figure 2-1a) to the stratosphere (see top of chart).

In short, by "floating" the base or bottom line up to $10,000, the reader or consumer is given a false sense of well-being; for the truth of the matter is that the salary increase is just 5 percent a year (or even a bit under this if compounding is taken into account). And that's hardly a rate to get really excited about in today's inflationary world.

A more truthful picture would require a base of zero income. Only then would the reader know just what the raise really meant to him. But too many chartists are reluctant to do this—not only because it does not serve their purpose, but also because such treatment makes for a very "flat" or uninteresting chart (see Figure 2-1b).

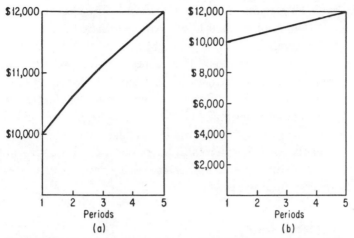

FIG. 2-1   The "zero distortion": (a) zero elimination, (b) a realistic scale.

Incidentally, it is the latter factor that prompts many of our newspapers and magazines to forget the zero line. The news media need "impact," and only by floating the base or eliminating the base zero line can they achieve this goal.

To be sure, some writers, editors, and statisticians have attempted to get around this floating base dilemma by including what is commonly called a "zero break"—that is, a little wavy line at the bottom of the chart to alert the reader to the fact that the base has been floated upward. But this is usually done in such an unobtrusive fashion that few readers (unless they're unusually astute) will notice the zero break insertion. Meantime, the chartist has covered himself. If anyone challenges him, he can point to the "fine print" warning at the bottom of the chart.

Politicians are particularly adept at what has now become known as the "zero ploy." If growth is the point in question, an incumbent will almost always forget the zero scale—in order to highlight the progress made during his time in office. On the other hand, if inflation is the bone of contention, then it is the challenger who will take the zero ploy approach, for only then can he make a relatively modest price advance during his opponent's term in office look like runaway inflation.

Before leaving this subject it might also be well to point out that elimination of the zero line is not always premeditated. Even reputable statistical agencies like the U.S. Bureau of Labor Statistics and the Census Bureau sometimes inadvertently start their charts at an other-than-zero level. In short, it is incumbent on every reader to look at the numbers which represent the lines as well as the lines themselves. It takes but a few seconds—and if such examination points up possible distortion, the time spent has certainly been worthwhile.

THE DISTORTED SCALE

In some cases forgetting to mention the omission of the zero line is only a starting point for the statistical manipulator. If he's unscrupulous enough, he can superimpose scale distortion on a chart already distorted by use of the above-mentioned zero ploy.

This action reinforces the effect of the zero ploy—namely, further hammering home or dramatizing the point that the writer or analyst wishes to make. Specifically, by changing the scale of values on the chart—by giving more weight to either vertical or horizontal moves— he can either (1) magnify or exaggerate existing movements or (2) de-emphasize such movement. Upshot: The reader is left with the distorted impression of what is actually happening.

Using a simple example—where income over a two-year period of time rises from $11,000 a year to $12,000 a year—all the chartist need do is change the dimensions of the chart, and the income gain looks either impressive or niggardly.

Assume first that the analyst who works for management wants to impress the worker with how fast his income is rising. His approach: Elongate the chart, giving more space to vertical (income) changes than to horizontal (time) changes. All this is dramatized in Figure 2-2a.

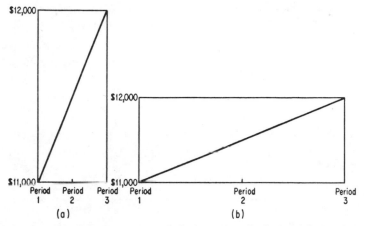

FIG. 2-2   Scale distortion: (a) vertical elongation, (b) horizontal stretch-out.

On the other hand, if the analyst is working for the union, he might be tempted to "stretch out" the chart, giving more space to periods of time and less to changes in income (see Figure 2-2*b*).

Both sections of Figure 2 tell the same story, if the reader would take the time to read the scales. But how many do? For the vast majority who only "look at the picture," the end result is a feeling that incomes have been skyrocketing (in the first case) or just slowly inching up (in the second case).

The lesson is clear: Elongate the vertical scale, and the amplitude of change is magnified; stretch out the horizontal scale, and the magnitude of change is dampened.

NONCOMPARABLE SCALES

There are also occasions where the chartist will fail to use comparable scales when depicting two different series side by side. His usual aim: To give a misleading impression of the relationship between the series being compared.

Again, an example can probably best illustrate the point. This time assume that income again went up from $11,000 to

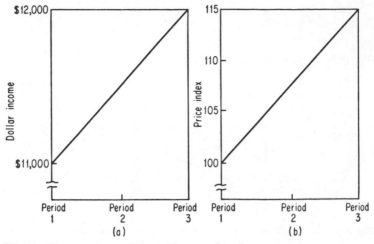

FIG. 2-3   Erroneous comparison: (a) income, (b) price.

$12,000 from Period 1 to Period 3. Also assume that prices went up from an index level of 100 to an index level of 115 over the same period. The chartist who again is working for management might plot them as shown in Figure 2-3.

A first glance by the layman might suggest that the rises have been comparable (the upward slope of the line in both cases is the same). Also, because the charts are of the same dimensions, the layman might well assume there has been no scale distortion. Upshot: The poor wage earner may leave with the feeling he has neither lost nor gained.

But closer examination will reveal otherwise. The rise in income over the period has been in the order of 9 percent. The rise in prices has been in the order of 15 percent. In other words, the income receiver has lost ground over the period under consideration. But he would never know it by looking at the charts in question.

The point to remember: When two series are being compared, the figures should be expressed in the same units—dollars, index numbers, tons, etc. If they're not, then the only meaningful graphic comparison requires the plotting of rates of change rather than absolute change, for the latter approach is an open invitation to distortion.

Incidentally, there's a way of plotting rates of change—and hence comparing two unrelated series—without the actual conversion into percentages. It's via the ratio chart, which is discussed below on page 30.

TWO-DIMENSIONAL DISTORTIONS

Up to now the discussion has been essentially one-dimensional —with charts stretched out either horizontally or vertically to achieve the desired result or implication. But this is hardly the only option open to the miscreant hell-bent on deceiving the reader.

Two-dimensional charts—where both horizontal and vertical scales play a role—can lead to equally disturbing misinterpretations. Some may find this hard to believe; for it can be argued that distortion in one direction (say, horizontally) would automatically be corrected by an offsetting distortion in the other direction (vertically). In other words, these people might argue that area charts—using both horizontal and vertical measurements—are a way out of the distortion box.

But seldom is this a way out. For the use of area representation can lead to even more serious interpretation errors— primarily because the eye cannot discriminate differences in area nearly as accurately as it can differences in height or width. Clearly the horizontal line A (2 inches long) is twice the length of line B (1 inch long).

Line A	Line B

And just as clearly vertical line A (2 inches long) is twice the length of vertical line B (1 inch long).

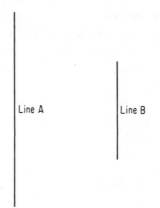

But combine the two into squares (a form of area diagram), and the 2:1 ratio no longer holds.

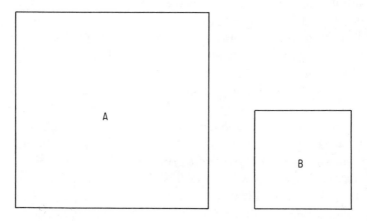

A little elementary arithmetic tells us that the area of square *A* is 2 x 2—or 4 square inches in all. On the other hand, the area of square *B* is only 1 square inch (i.e., 1 x 1 = 1). But it would take a really astute observer to look at the picture of the two area diagrams above and know that *A* represented a magnitude four times as great as *B*. Try it out on your friends. Chances are you'll get answers ranging from 2:1 to near 3:1.

In short, beware of area diagrams—charts which purport to measure differences in magnitude by differences in area. It's the rare person who can even approximate the difference between such area diagrams.

Then why use such diagrams? Again the reasons are virtually the same as noted for one-dimensional charts: distortion and impact. If you want to play down differences, area charts are just what the doctor ordered. Then, too, if you want impact, the area diagram is hard to beat. Thus two different-sized dollar bills to measure profit change—or two different-sized airplanes to measure military preparedness—are a lot more eye-catching than simple one-dimensional lines.

But again the question has to be asked: Do you want to be entertained or informed? If the answer is the latter, then use of the area approach is extremely dubious. But there is a way out in this case. Label each area chart with its numerical equivalent. This kills two birds with one stone: impact is maintained, and at the same time the possibility of misinterpretation is cut down considerably.

MEASURING GROWTH

Even the seemingly simple concept of growth can lead to serious misunderstandings when numbers are transferred onto charts. The whole problem can probably best be illustrated via a simple example. Brand A's sales go up from $1 billion in year 1 to $5 billion in year 2—and to $25 billion in year 3.

Plot this on a traditional or arithmetic grid and the picture illustrated in Figure 2-4a emerges. Plot the same numbers in terms of rate of growth (a fivefold increase in each of the years involved), and the picture illustrated in Figure 2-4b emerges.

The charts look a lot different. But is one wrong and the other right? Here we have to equivocate on the answer. If we are measuring dollar growth, then Figure 2-4a is the correct one.

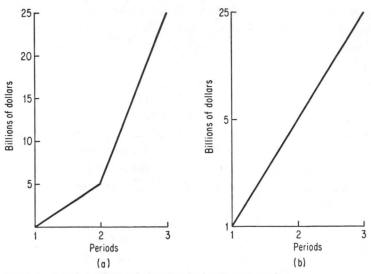

FIG. 2-4 Brand A sales: (a) arithmetic scale, (b) ratio scale.

But if we are interested in measuring the rate of growth, then Figure 2-4b is the pertinent one.

It all boils down to what concept we are interested in. If cash flow or profits is the crucial issue, then Figure 2-4a might be the more useful one. On the other hand, if we are interested in relative performance over a period of years, then clearly Figure 2-4b is the relevant chart.

Statisticians have a name for both charts. The first is an *arithmetic chart*—equal dollar changes showing up as equal distances on the chart. The second is known as a *ratio* or *log chart*—equal percent changes showing up as equal distances on the chart.

But a word of caution on the more popular of the two—the arithmetic grid. Unless one realizes what is being shown, serious distortion can result. Thus even in the short period of time shown in Figure 2-4, if we concentrated exclusively on Figure 2-4a, we might be led to believe we were performing a lot better in the latter period. But the truth of the matter is that there exists a stable or straight-line rate of increase over the entire period in question.

In short, relatively speaking, our performance was about the same in both periods being monitored. In any case, the rules on when to use each type of chart are simple enough. If you are interested in absolute growth, then the arithmetic grid (Figure 2-4a) is the one called for. On the other hand, if the rate of growth is what you are after, then the ratio grid (Figure 2-4b) is the appropriate one.

A few further words on when the ratio grid is the more useful one may be in order. Specifically, concentrate on the ratio chart

■ When measurements are strung out over long periods of time. That's because dollar increases become irrelevant over such long spans. Clearly the dollar increase in GNP today is larger than it was in 1800. So any charting of the historical path of GNP over such a long span must be in terms of growth rates. Ergo, the ratio chart is the appropriate one.

■ When comparing series of different magnitudes. Say you want to compare the profit performance of General Motors with that of the much smaller American Motors. It would make little sense to look at dollar figures because of the tremendous gap in magnitudes. But compare the rates of change between the two companies (via a ratio-type chart), and your chart can prove surprisingly useful.

On the latter score, the ratio chart can also prevent misinterpretation when two series are being compared one of which is far larger than the other. For example, plot over a period of time the purchases of a millionaire against those of a poverty-level family on an arithmetic grid. The fluctuations of the poor family will seem inconsequential. That's because the scale (to encompass both income levels) must be calibrated in thousands of dollars—far too wide to dramatize even an appreciable change in the poor family's purchasing habits.

This might then result in the erroneous conclusion that the

poor family's spending habits are relatively stable—or, conversely, that the millionaire's purchasing patterns are extremely erratic.

Again, the use of a ratio approach could prevent this impression because both families are compared on the basis of rate of change. If both move up or down by the same percentage, this shows up as an equal space shift on the chart. Put another way, the ratio grid facilitates comparisons of performance when the magnitudes being compared are of uneven size.

Another ratio chart plus: Percentage changes can be obtained directly from the ratio chart without any arithmetic calculations, for it can be shown that the steepness of the plotted line at any point on a ratio chart represents nothing more than rate of change.

But for those who have an aversion to working with ratio charts, mathematicians have provided an out. Instead of plotting the absolute figures, the rates of change are plotted on an arithmetic grid—and the same basic effect is achieved. The calculations may be a bit more laborious, but, on the other hand, an arithmetic grid is less likely to frighten off some of the less sophisticated readers.

THE PARTIAL PICTURE

Actual wrongdoing need not be limited to the use of an irrelevant or distorted chart. Just as in the case of pure numbers, the statistical deceiver can choose to present only that part of the picture that will establish his thesis. In a way, the problem is compounded when graphics are involved—for readers will tend to focus a lot more closely on charts than on rows and rows of dull figures.

In other cases, the distorter can follow the letter of the law— charting all the relevant statistics—and still leave the reader with a misleading impression. Specifically, by judicious use of color, overlays, type of chart, sequence of charts, etc., one can play down one part of the overall picture while playing up the other. Again, technically speaking, the chartist has done no wrong. He has given you everything. If you have focused on only one aspect of the situation, that's your fault, not his. Thus, if sales are up and profits are down, the annual report might dress up the sales chart while making the earnings chart nondescript or extremely dull.

Bad as the above distorter is, he is not nearly as serious a wrongdoer as those who choose to present only part of the pertinent information on a graphic basis. Advertisers, of course, are past masters at this type of deception. A seller of condominiums, for example, will chart only the rock bottom cost—but conveniently forget to add on all the monthly extras. Similarly, a company that has been gaining in terms of overall sales but losing in terms of share will play up the former and forget the latter. Unfortunately the latter type of distortion is extremely hard to spot, for unless one knows the industry, there's no way of knowing what pertinent information may have been omitted.

THREE

THE PERCENTAGE PLOY

You can't live with 'em, you can't live without 'em.

No, this isn't another male chauvinistic remark directed against the fairer sex. Rather, it's about the only realistic description of percentages that can be given these days.

On the one hand, much of our statistically oriented society is dependent on expressing one number in terms of another. Yet it is equally true that percentages are the lifeblood of the statistical charlatan. Without the opportunity to compare one number with another, a good portion of today's distortions and outright lies would all but disappear.

This is in no way meant to demean the legitimate use of percentages. Indeed, pick up your daily newspaper and you'll agree that it's the rare story that will not have some reference to a percentage. Index numbers, interest rates, discounts, markups, rates of growth—all these and a host of other commonly accepted yardsticks are simply nothing more than variants of a percentage.

Another percentage plus: Its use tends to simplify analytic thinking by converting large complex numbers into smaller, simpler ones. Thus the rise in our population from 200 million to 202 million is more readily grasped by stating that our population rose by 1 percent. Finally, in many ways the percentage approach is often little more than just good old common sense. Thus, stating that a family earned $1,000 more this year than last year is of little value. If it's the income of a millionaire that's involved, the increase is negligible. On the other hand, if it's the factory hand down the street whose income is being analyzed, the $1,000 increase is clearly significant. But in either case, it is the percentage (comparing the increase to a base-period yardstick) that gives meaning to the figure.

Obviously, then, percentages are indispensable. Nevertheless, difficulties do arise—primarily because this handy tool is so easy to manipulate. In some cases, for example, percentage notation is used not so much to impart information as to give numbers an aura of respectability. There's something believable in the statement that 62 percent of the people prefer brand A, even though the figure may be completely irrelevant or fallacious. Somehow, a percentage smacks of research, deep study, and careful analytical thought—things which an absolute number doesn't always connote.

In some extreme cases the use of percentages amounts to little more than playing the numbers game. Take an obvious example culled from the political arena. Consider candidate A, who is recognized as the front runner for the presidential nomination. A pre-primary poll is then taken which shows that candidate A leads the field with about 75 percent of the vote.

One would expect candidate A supporters to be overjoyed.

Quite contrary. More likely than not, they'll say the figure is absurd—an "impossible standard."

And their attitude is understandable. One plays the "primary" game by poor-mouthing one's expected percentage of the vote while inflating the percentage of one's opponent. The aim is to impress people by the actual percentage garnered on primary day. If you convince them that you were expected to get only 60 percent of the vote and wind up with 75 percent, you are a hero. On the other hand, if they believe that you should have polled 75 percent and you wind up with only 65 percent, then you may well be considered a political has-been.

This percentage game, then, is obviously a handicap for front runners, but can be a real launching pad for unknowns. If the public believes that a dark horse can expect 8 percent of the vote, and then this outsider corrals 15 percent, he's off and running—with his supporters touting the fact that he doubled his expected percentage but conveniently forgetting to point out that his front-running opponent ran off with the majority of the vote.

The list of potential distortions involving percentages is so long that the remainder of this chapter and all of the next will be devoted to this one area. It is hoped this will serve to put all the relevant percentage caveats into proper perspective.

A HANDY SMOKE SCREEN

Many times percentages, since they are so widely recognized and accepted, are used as a tool to obscure the facts—by either hiding the truth or giving a misleading impression. Aside from the presidential primary example noted above, here's a rundown on a few of the more popular of these ploys:

1. *The less-than-full picture.* Quote a percentage, and the gullible public is more than willing to accept it as revealed wisdom even though it scarcely sheds light on the subject under discussion. How often, for example, have you heard an industrial-

ist brag that, say, most Americans have a stake in running American industry. To buttress this questionable assumption, he will point out that nearly one out of every three American families owns stock in a United States corporation. This sounds quite impressive. But it is only part of the story—for the fact is that while one out of every three Americans does own some stock, the number of shares he owns is so small as to make the original assertion ludicrous.

If the whole truth were to be given, it would have to include the fact that only about 1 percent of United States families own about two-thirds of all the stock. In short, the meaningful percent as far as control is concerned is not the number of families who own securities but rather the concentration of holdings within the ownership group.

Another even more blatant example of the partial information approach: the recent quoting of figures purporting to show that the middle class is the group that benefits most from what are commonly called "tax loopholes." What's left out of such analyses is how many people there are in the various income groups being compared—a bit of information that is crucial to the resolution of the tax loophole-benefit controversy.

EXAMPLE: Those fighting against the closing of loopholes back in 1972 were quick to state that persons in the $10,000–$15,000 income group realized a total of $642 million in tax savings the previous year from the section of the tax law that permits deductions of property taxes. Persons in the over-$100,000 bracket, loophole proponents hastened to add, realized only about one-fifth as much, or $137 million.

But at that time there were more than 14 million individuals and families in the $10,000–$15,000 bracket—about 20 percent of all taxpayers—while there were only about 78,000—or less than two-tenths of 1 percent of the total—in the over-$100,000 category.

Thus, when the benefits of the property-tax deduction are calculated on a per-taxpayer basis, the savings for those in the $10,000–$15,000 bracket averaged $45.52 a year, whereas in the over-$100,000 bracket the savings averaged $1,758.66.

The disparity created by the special tax treatment of capital gains was even greater at that time. The middle-income taxpayers in the $10,000–

$15,000 class realized an average saving of $16.31 a year from this tax-law provision, but those in the over-$100,000 bracket realized an average of $38,125.80.

The above calculations were worked up by the Tax Reform Research Group, an organization created by Ralph Nader. The organization accused vested interests of "presenting deliberately misleading statistics to the Congress and the American public in attempt to undermine the growing pressure for tax reform."

This same study by the tax reform group showed that even tax benefits that have been kept in the law primarily out of concern for persons of average income benefit the wealthy more. An example is the tax-free status of the first $100 of dividends received each year by an individual. According to the reformers, for persons in the $10,000–$15,000 income bracket the average tax saving created by this provision was $3.90, whereas in the over-$100,000 bracket the average saving was $115.53.

2. *The irrelevant percentage.* Landlords generally are past masters in the use of this ploy. In the early 1970s they were consistently citing 10 percent and 15 percent increases in fuel and janitorial service costs to justify rent increases of the same magnitude. What these house owners conveniently forgot to stress was the fact that interest payments—their biggest single cost—had not gone up. In short, landlords were quoting skyrocketing variable costs but ignoring relatively stable fixed costs.

This isn't too surprising, for a look at the rise in total costs could scarcely be deemed alarming enough to justify the big rent increases being demanded. Take the experience of New York City's Rent Stabilization Board. This watchdog agency, alarmed by the rent increases being asked, did a comprehensive study. Not surprisingly, the group found that overall landlord costs were rising only about 4 percent a year, and hence they refused to okay anything bigger in the way of annual rent increases.

The manufacturer who cites his need to raise prices, say 5 percent because his labor costs have gone up by this amount, is guilty of the same basic distortion: projecting the percent change in one cost onto all costs. More will be said about this latter distortion in Chapter 10.

At other times, the attempt to relate a percentage to a particular thesis or argument borders on the ludicrous. Thus, the author, when first hearing the TV commercial detailed below, refused to believe his ears. Only after hearing it repeated a second and a third time did he realize there was a method to the madness.

Specifically, a few years back, a large commercial bank was advertising the fact that depositors could bank by mail. Assuming a 50-cent bus fare both to and from the bank and 15 trips a year, the financial institution stressed the point that a depositor could save $15 a year. So far so good—if you're willing to accept the idea that the 15 trips in question were for the sole purpose of banking. But not content with even this questionable assumption, the bank wound up with this fantastic non sequitur: namely, that the $15 savings was the equivalent of 1 percent of $1,500. What this last statement had to do with the possible savings an individual depositor might realize from banking by mail was anyone's guess.

Obviously there was no connection, except for the fact that the bank wanted to leave the listener with the vague feeling that substantial sums of money were involved. After a number of irate phone calls, the bank came up with the lame excuse that two commercials had been inadvertently mixed up—a pretty hard thing to swallow from an institution that had heretofore prided itself on its precision and accuracy.

3. *The not-so-relevant percentage.* In many instances, there's a tendency to assume that a percentage designed to measure one phenomenon can be used to measure a seemingly closely related phenomenon. But this assumption is often unwarranted.

This hypothetical example can perhaps best illustrate the pitfall here. A labor union manages to win 80 percent of its original demands for a pay rise from $4 to $4.50 per hour. In other words, a pact is finally inked calling for $4.40 per hour.

Next year the union, its confidence bolstered by the previous year's success, asked for the impossible: a hefty $2-per-hour boost (to $6.40 per hour). But in the end the union has to tone down this ridiculously high figure to 40 cents per hour, thus winning only 20 percent of its original demands.

Someone reading these percentages—namely, that the union won 80 percent of its demands the first year and only 20 percent the second—might infer that wage hikes were moderating or that union strength was on the wane. But this is, of course, sheer nonsense, for the union managed to win the same 40-cents-per-hour increase in both years.

The moral is clear: Always use a number only for what it was designed to measure. In this case, there's absolutely no justification for equating union demands versus union achievement with actual wage rate increases, for the two concepts are not necessarily correlated.

A similar pitfall involves the equating of consumer desires with actual consumer purchases. They're seldom the same. To take an extreme example: Some 50 percent of the families responding to a consumer survey might indicate they'd like to own a Cadillac or a Lincoln. But that's not the same thing as saying that 50 percent of families buying cars will actually purchase these expensive models. Chances are that income realities will force most of these potential car buyers to settle for a Chevy or a Ford.

4. *The magnitude of the base.* Percentage changes are often extremely sensitive to the number from which the change is being measured. Other things being equal, a small number is likely to change by a bigger percentage than a large number. If you start with the number "1," then a 1-unit increase is the equivalent

of a 100 percent advance. But that same 1-unit increase amounts to only 1 percent if the base number should be "100."

Distortions arising from a small base often arise when a new industry is being developed. If an African nation, for example, starts an auto industry from scratch, its output goes up by an infinite amount during the first year—and probably several hundred percent during each of the next few years. On the other hand, a nation like the United States, with a mature auto industry, would be hard put to raise auto output by, say, more than 10 percent—not only because the amount of capital investment would be prohibitively high, but also because demand is already close to the saturation point.

A similar misunderstanding crops up when comparing the rates of growth of mature nations with those of underdeveloped nations. A poor agrarian economy, by means of the installation of a few factories could easily increase its GNP by, say, 10 percent or 15 percent. On the other hand, the United States or any Western European nation would be hard put to rack up this kind of gain. Even Japan, which has been growing at a better than 10 percent annual rate, is finding this out. As this Far Eastern nation becomes richer, big year-to-year gains are harder and harder to come by—so much so that even optimistic Japanese politicians have had to tone down their growth forecast over the next decade to around the 7 percent to 8 percent annual level.

Note, too, that while some of these poorer nations have shown strong percentage gains, growth in actual terms has declined vis-à-vis the United States.

Thus, because the United States base is so much larger than an African country's, a 5 percent GNP growth here will yield a greater physical increase than even a 25 percent or 30 percent increase in such an underdeveloped nation.

A simple example: We increase our widget output from 1 million to 1.1 million (a 10 percent gain). An underdeveloped country increases its widgets from 100,000 to 150,000 (a 50 percent gain). In absolute terms, we have increased our widget lead from 900,000 units (1 million less 100,000) to 950,000 (1.1 million less 150,000).

This is not an unimportant development. The fact that an absolute gap can be growing when a percentage gap is being narrowed is often quite significant. This is particularly true when prices are involved.

Again a simple example can best illustrate the principle. If we sell widgets for $5 per unit because of our high productivity, while our competitor sells them for $20 because he uses much more hand labor, then we can afford to raise our prices much more than our competitor and come up with an even bigger absolute advantage than heretofore. Assume we boost prices by 10 percent while our competitor boosts prices by 5 percent. Our new price is then $5.50. Our competitor's new price is $21. The original difference was $15 ($20 less $5). The new difference is $15.50 ($21 less $5.50).

While the example is exaggerated, the principle is important because it is the dollars-and-cents advantage rather than the percent increase that ultimately determines which seller gets the order.

5. *Obscuring the sample size.* One often used trick is to use a percentage to hide the fact that an extremely small sample has been used. How many times have you read a figure that suggests that 80 percent of the people prefer brand A? Does this 80 percent represent 4 out of 5 surveyed people, 40 out of 50 people, 400 out of 500 people—or what? Nobody knows. But chances are that if the number sampled isn't spelled out, it is small. The trouble, however, is that if only five people were surveyed, the results would be entirely unreliable from a statistical point of view. (More about what constitutes an adequate sample in Chapter 7.)

In any case, whenever percent figures are used, it is always a good idea to ask about the size of the sample. Most surveyors, if they're anywhere near reputable, will be more than willing to supply the information. (If they don't, then the sample should be regarded as suspect.) Once this figure is known, it is also a good idea to inquire into the actual sampling technique (see Chapter 7 again). With these two additional bits of information (sample size and type of sample), it is then possible to gauge the reliability of the sample. And if you can't do the calculating, anyone with a one-year course in statistical sampling can probably do it for you in a matter of minutes.

PERCENTAGES AND TAX RATES

In many cases, published percentages bear little resemblance to actual percentages. Thus much has been made of the fact that the percentage of income paid to Uncle Sam rises with the level of income, and that when income reaches high levels, the income tax rate can be in excess of 50 percent. In short, there's a widely held belief that we have a highly "progressive" tax structure.

Nothing could be further from the truth. It's no secret, for example, that higher-income brackets are often not subject to general tax rates. Loopholes—including capital gains and other tax shelters—in almost all cases reduce the effective rate of tax paid by the income recipient.

If there's any doubt of the inequities, consider these 1969–1970 statistics:

> Some 112 Americans with annual incomes of more than $200,000 in 1970 paid federal income taxes of zero—despite the much heralded tax reform of 1969 and the new minimum income tax of 10 percent imposed at that time. Corporations paid an estimated average effective tax rate of 37 percent of profits in 1969, according to a recent study. Moreover, the top 100 companies managed to reduce the toll to 26.9 percent, a far cry from the 48 percent tax legislated by Congress.

Further support on the lack of progressivity comes from Roger Herriot and Herman P. Miller of the Census Bureau in their landmark 1968 study on tax distribution. Even if you add in various benefits made via government transfer payments (Social Security and the like), the study finds that, for the bulk of Americans, the tax structure is clearly not very progressive. Details are shown in Table 3-1 below.

Note the narrow tax-rate range of 25 percent to 31 percent for income of less than $2,000 to as high as $25,000. This can hardly be viewed as the basis for substantially progressive tax structure. For within these groups at that time were 96 percent of all families and unrelated individuals, and their total income before taxes and transfer payments accounted for more than 80 percent of all income.

Indeed, for most Americans in 1968, taxation was proportional: that is, taxes took equal proportions through a wide range of incomes. Note that tax rates through the $6,000–$25,000 levels show an even smaller spread of only 1.2 percentage points. For over 60 percent of us—a group that in 1968 accounted for 72 percent of all income—it seems that progressive taxes like the income tax are counterbalanced by regressive taxes like those on property—leaving us, like it or not, with proportional taxation.

Table 3-1   INCOME VERSUS TAX PAID

Income Group	Tax Rate
Under $2,000	25.6%
$2,000–4,000	24.7
$4,000–6,000	27.9
$6,000–8,000	30.1
$8,000–10,000	29.9
$10,000–15,000	30.9
$15,000–25,000	31.1
$25,000–50,000	33.6
$50,000 and up	46.6
Total	31.6

Another misconception about tax percentages: Talk would have us believe that federal rates were falling in the late 1960s and early 1970s. But closer examination would seem to suggest otherwise.

It all stems from the fact that there's a tax on inflation. Specifically, the effect of inflation has been to raise money incomes and, because of the graduated nature of the tax tables, to increase the tax burden falling on the mass of wage earners, without any change in rates.

EXAMPLE: If a man earned $10,000 in 1965 and his salary increased by 23 percent between 1965 and 1970, his gross income would have just kept pace with the 23 percent rise in the cost of living over this five-year period. But if he took the standard deduction and claimed four personal exemptions, his tax payment would have risen from $1,114 in 1965 to $1,556 in 1970—or by nearly 40 percent.

Viewed from another perspective, the taxpayer who in 1965 was paying 11.14 percent on every dollar of income would be paying 12.65 percent in 1970 ($1,556 tax on $12,300 income). That's a 14 percent increase in the effective tax rate (12.65 percent versus 11.14 percent).

The net result of all this would be a drop in the taxpayer's real standard of living. Specifically, his old take-home pay of $8,886 ($10,000 less the $1,114 paid out in income taxes) has given way to a new take-home pay of $10,744 ($12,300 less $1,556). When you correct this new $10,744 take-home pay for a 23 percent rise in the cost of living, it drops down to $8,735—actually under the 1965 purchasing power figure of $8,886. In short, the taxpayer in terms of real purchasing power would be 1.7 percent worse off in 1970 than he was in 1965 ($8,735 versus $8,886).

HIDING THE TRUE INTEREST RATE

Differences between effective and listed rates, of course, are by no means limited to the taxation area. Interest rates are also prone to the nominal-effective rate distortion—so much so that a few years ago the government decided to pass a comprehensive truth-in-lending law. The deception is quite evident if one would take the trouble to do a few simple calculations. Assume you want to borrow $1,000 for a year from a bank. The bank agrees to a 6 percent loan, but takes out or *discounts* the interest in advance, so you wind up with only $940. In essence, the bank is charging you $60 for the privilege of borrowing only $940. In short, your true interest rate is something over 6 percent.

Then if the bank really wants to get cute, it asks for repayment in monthly installments. This again raises the interest rate, because you don't even have full use of the $940 for the entire year. The exact interest rates and how they are distorted will be discussed more fully in the following chapter.

Nor is a businessman always immune from the effective-nominal interest rate distortion. That's because a bank normally requires a firm to keep a given amount of a loan, known as a *compensating balance,* on deposit at all times. Say your business gets a $1,000 loan at 7 percent, with a proviso that 20 percent of the loan be kept on deposit. This effectively gives you the use of only $800—in return for the payment of $70 in interest. That adds up to an effective rate of close to 9 percent.

SHIFTING THE BASE

One of the most common errors involving the use of percentages is the assumption that a given percentage decline and a subsequent percentage advance will cancel out. But this is sheer nonsense. Take the following exaggerated example, where sales decline from 1 million units to 500,000 units (a 50 percent decline). This is then followed by a 50 percent advance. But the latter hardly brings us back to 1 million units; for a 50 percent increase on 500,000 units suggests a sales bounceback to only 750,000—a hefty 250,000 units under the original sales level.

Why don't the two 50 percents cancel out? It's all a question of base. The 50 percent decline applies to a base of 1 million, but the subsequent 50 percent rise applies to a base of only 500,000. Since the decline is from the larger 1 million base, the net result is a smaller figure after the decline and subsequent recovery.

To be sure, when the declines and recoveries are in the area of 1 percent or 2 percent, the discrepancy is small because the base change is small. But in extremely volatile areas where swings of 15 percent or more are not uncommon, this base-shifting distortion can't be ignored.

There are other equally serious types of base-shifting distortions, too. One of the most common of these is: calculating profit margins from the largest possible base to make profits seem as small as possible. It should come as no surprise, for example, that retailers figure margins as a percent of the sales dollar rather than as a percent of cost. If they chose the more tenable latter approach (profit should always be figured as a percent of money invested), they would end up with higher margins—ones that might be more difficult to explain to customers. This trek toward the use of a base yielding the lowest margin has become virtually standard operating procedure, so that retail markups are now automatically assumed to be computed on a sales rather than a cost base.

A somewhat similar approach is used by manufacturers who prefer to quote profits as a percent of sales rather than as a percent of investment or stockholders' equity. Again the reasoning is the same: margins are almost always smaller when using the sales gauge—and hence are preferred by firms wishing to establish the need for still higher margins.

But perhaps the most blatant type of base-changing distortions are those involving *index numbers* (a number expressed as a percent of a base year). Almost everybody is guilty here. If you want to prove your approach to crime prevention is successful, you compare current crime to some previous peak rather than to a more normal period. Similarly, labor will compare current real purchasing power with a previous high (to play down the gain), while their management counterparts will compare it with a particular low period to emphasize how much it has risen.

We all have done the same thing with prices, unemployment, or any number of other economic or social measures. The tack is always the same. Compare with a high base if you want to play down a rise—and with a low base if you want to play it up. Declines, of course, would work the other way around. If a store is running a sale, it compares prices with a previous high—to point up the extent of the sale. On the other hand, the incumbent politician is likely to compare a fall-off in industrial activity with a previous recession or low point—to de-emphasize the seriousness of the situation that developed when he was in a position of power.

THE CHANGING PERSPECTIVE

Viewing percentage changes from several different bases is often another way of parrying the possible bad press one might get from change. Prices are tailor-made for this type of approach. Thus if a company boosts prices by 5 percent it may come out with a statement to the effect that the increase is less than it seems—either because (1) the quality of the product has been upgraded or (2) "extras" which are usually tagged onto the base price haven't been raised.

Auto companies regularly take this approach. Thus in 1970 when they raised prices, they diminished the impact by making adjustments for quality and the fact that extras (power steering, air conditioning, etc.) had not been raised. On the latter score they pointed out that these items are normally purchased with a car, so they should rightfully be averaged into any overall car price increase. Detroit's basic aim: to convince the consumer that things weren't quite as bleak as they might have seemed at first blush.

Here's how General Motors proceeded that year: The company started with an announced 3.9 percent increase on the basic car. But then they watered this boost down to 2.1 percent by the above-noted approach. In short, adjustments for quality and extras helped cut the increase down to nearly half the announced figure. The actual figures are detailed in Figure 3-1 below.

All the above is in no way meant to imply that such a procedure was deceptive—for the deductions are certainly justifiable. Indeed, subsequent calculations by government price watchers vindicated the GM approach—with the increase in the price of 1970 model cars eventually showing up as a 2 percent rise in Uncle Sam's wholesale price index—just about in line with Detroit's own calculations. (More about how the quality adjustment is calculated in Chapter 12.)

Nor are the auto companies the only ones who make use of this approach. Machine tool people in selling more expensive numerically controlled equipment constantly downplay the higher cost of such machinery by pointing out that the quality improvement (in the form of higher productivity) more than offsets the apparent higher cost.

One problem with this approach, of course, is that not everybody benefits from such improvement. Thus the consumer who buys a car for basic transportation—and doesn't care about safety, pollution control, or expensive extras—winds up paying the full 3.9 percent price increase. For him, then, the 2.1 percent increase ultimately arrived at by Detroit is irrelevant. Similarly, the manufacturer who has to buy a more expensive machine,

Announced increase in basic car price: 3.9 %

FIG. 3-1   Components of an apparent price rise.

even though he can't make use of its additional productivity, ends up paying a lot more than is suggested by company adjustments to bring the price down.

In short, the adjustments are legitimate. But they refer to the average buyer. If you're not the average buyer, then the price increase you end up paying is the original boost—and not the adjusted one suggested by the company and usually rubber-stamped by Uncle Sam's price monitors.

Viewing a base from different perspectives can sometimes also lead to apparent paradoxes. Thus when the United States dollar was devalued in early 1973, American citizens read that the net effect would be to increase the cost of an import by about 11 percent. On the other hand, overseas buyers of our goods were told they could expect only a 10 percent reduction.

An example can perhaps best illustrate this 1 percentage point discrepancy. Take an import from Germany, where the old "3.22 marks equal a dollar" relationship was changed to a "2.9 marks equal a dollar" relationship. Assume next you are importing a bottle of German wine, for which the German vintner wants 3.22 marks. Before devaluation you exchanged your dollar for 3.22 marks, which enabled you to purchase the bottle of wine. But after devaluation you would have to pay out approximately $1.11 to obtain the needed 3.22 marks. (You would get 2.9 marks for one dollar and 0.32 marks for an added 11 cents.) Together this would total up to 3.22 marks, the amount needed to purchase the wine.

Next, consider the German who had been buying a $1 widget from a United States seller. Pre-devaluation, he would turn in his 3.22 marks to get his dollar with which to buy the widget. Post-devaluation, he would turn in only 2.9 marks to get his dollar. The savings: 0.32/3.22 marks—or roughly 10 percent.

In short, one man's 11 percent loss is another man's 10 percent gain. The apparent paradox again stems from the different ways of looking at the change. In the case of the United States importer, the change is being figured in terms of how many marks a dollar can buy. But the German importer is figuring in terms of how many dollars his marks can buy. Different viewpoints and, hence, different percentages.

SHARE ISN'T EVERYTHING

A one-quarter share of a 12-inch pie is a lot more satisfying than a one-third share of an 8-inch pie. This is pretty obvious when you're chomping away at your favorite pastry—but not quite as obvious when the share concept is applied to other areas.

Yet what is true in a gastronomic sense is equally true in an economic or monetary sense. Thus, a rising share of sales in a declining market can often be misleadingly optimistic. When everybody was shifting away from black and white to color TV, such an increase meant little to the company that continued to specialize in black and white sets. Its share of the market might have been rising, but its absolute sales were falling—and profits,

"I have good and bad news. Our new item has broken all sales records, but due to a cost-accounting error, we've lost fifty cents on each item."

after all is said and done, are dependent upon sales volume and not market share.

Even when dollar sales are expanding, an increasing share is not always consistent with improving profits. If the price is declining, share means little when the firm is producing at a loss or at very little profit. The old anecdote comes to mind about the firm that cut prices below its actual costs to force out its competitors. When asked how the firm could survive, the answer was, Volume.

The story—as well as the cartoon on page 51—is funny because it points up the incongruity of setting one's goal exclusively in terms of share. The point, of course, is that growth in volume alone is a meaningless measure. It is important only to the extent that it affects profits—the raison d'être of all private enterprise.

On the other side of the coin a maintained share, or even a declining share, might not be as catastrophic as share percentage alone would seem to indicate—particularly if growth is attracting many new suppliers into the market. Under these conditions absolute sales would still be increasing at a fast clip, and it would be the height of fantasy to assume that such prosperity conditions would not attract competitors.

Another "share" fallacy: the tendency of some to assume that an increasing supply of goods or services will automatically result in a diminishing share of sales for each individual supplier. The fact is that if increased supplies are made available, demand may increase commensurately. Consider taxi service. This badly maligned profession has consistently opposed an increase in the number of cabs, fearing that it would lead to a drop-off in business for each individual driver.

But moment's thought might suggest otherwise. An increasing supply of cabs—particularly during periods of inclement weather—may well result in an increase in taxi usage, simply because additional cabs would then be available. Recent experience in New York with nonmetered taxis would seem to bear this out. Thousands of such additional cabs were put on the street. Contrary to metered-taxi fears, overall business stayed high—despite a stiff rise in all fares that took place at that time.

THE DECELERATION DECEPTION

Percentages can often be used to give the impression that things are getting better when in fact they are actually getting worse. Usually some long-festering problem is involved. It can be any one of the socioeconomic headaches we face today—or even such a mundane business dilemma as a loss in sales or profits.

But whatever the subject, the technique is always the same: Point out that things aren't getting worse at quite the rate they were a month or a year ago. Thus, Washington spokesmen will say that the rate of inflation this month was 3 percent instead of the 4 percent figure reported the previous month. Sounds like improvement. But is it really? Such a statement ignores the fact that in absolute terms the situation is still deteriorating. You are paying more today than you did yesterday for your goods and services purchases.

Reports on the incidence of crime can be equally misleading. In 1972, a Justice Department report stated that serious crime in 1971 "registered the smallest rate of increase in six years"—7 percent. This marked the third consecutive year that a "tapering off has been reported in the growth of crime," the release continued. Figures were cited to show that the annual rate of increase in the three years before the Nixon administration took office were 11 percent, 16 percent, and 17 percent, and the figures for the first three Nixon years were 12 percent, 11 percent, and 7 percent.

"Smallest" and tapering off" sound as though the administration, which came into office brandishing a tough law-and-order policy, had licked the criminal element. One radio announcer on a Washington, D.C., station announced with what seemed like awe that "crime in 1971 rose only 7 percent." The Attorney General, meantime, boasted that the United States was moving from a crime increase to an actual "crime decrease."

The point forgotten was that the crime rate was still climbing. In 1968 there were 4,477,200 serious crimes committed in this country. The figure went up in 1969, in 1970, and again in 1971. In 1971 there were 5,995,200 serious crimes committed. That's 1,518,000 more than in 1968, or a 33.9 percent increase over the three-year period.

Forgotten, too, was the fact that the increase in population over the same period was only about 5 percent—far less than the one-third increase in crime. In other words, the chance of an individual citizen becoming the victim of a serious crime had increased substantially over the three-year period in question. To boast about improvement under these circumstances would seem to be the height of deception.

CONFUSING PERCENTAGES AND PERCENTAGE POINTS

Read your daily newspaper and you find it is often difficult to learn just how much a price or production figure went up or down. The writer will say the index went up 1 percent—but is he referring to the real percent increase or the number of index points? Nine times out of ten the reader will never find out unless he gets access to the report being referred to. And it can make quite a difference—particularly when the magnitude of the index number is large. Thus some indexes, where the base hasn't been changed in decades, may be hovering around the "500" mark. If so, a 1-index-point change would be the equivalent of only a one-fifth of 1 percent real change (501/500) in the variable being measured.

The point to remember: The difference between index point and percent change tends to grow as the magnitude of the index number grows. But even when an index is in the 150 range—as many now are—the difference can be significant. Thus when the index rises to 151, it is a 1 index-point advance but only a two-thirds of 1 percent increase.

This is one reason why statisticians prefer periodic updates of index-number base periods. Shifting the base to a more current period brings the actual reading back closer to 100—and thus reduces the possibilities of misunderstandings between percent and index-point changes (see Chapter 11).

All the above is not meant to denigrate some legitimate uses of percentage points—particularly where opinion polls are involved. There is nothing nefarious in talking, say, about a 4-point spread between the 52 percent who prefer brand A and the 48 percent who opt for brand B. But even here differences are not quite as solid as they seem. Thus in the above example, a 2 percent swing could completely wipe out the 4-point spread—and put both brands on an even keel. Incidentally, that's what makes political poll taking so risky.

Chance factors further complicate such opinion samples. As noted in Chapter 7, chance variation occurs in all samples. Hopefully this latter type of variability can be adequately appraised by using statistical probability—provided, of course, one takes the time to do the necessary calculations.

But, unfortunately, probability theory cannot be applied to the opinion-shift part of any variability. This portion of sampling error is simply not predictable.

THE CHANGING SIGNIFICANCE
OF PERCENTAGES

Too often there is a tendency to assume that if a given percent meant something in one time period it will continue to mean the same thing in all subsequent time periods. But this assumption is not always warranted. This is so even when the data are completely free of any statistical errors and/or distortions. These problems of interpretation and evaluation usually occur because of the ever-changing structure of our socioeconomic system.

Unemployment is certainly a case in point. A decade or two ago, a 6 percent unemployment figure would have been accepted as unavoidable—for it was then recognized as the norm or average. But not so any more. Times have changed. Increasing social unrest, combined with government intervention, has dropped the norm down to near 4 percent. Indeed, full employment has now become synonymous with only a 4 percent jobless rate. And few would question the decline. Indeed, there are some who feel it is still too high—that a 2 percent or 3 percent figure would be a more realistic jobless goal.

The acceptable unemployment rate has also been driven down by our changed attitudes on just what constitutes unemployment. On a statistical basis, all jobs count: part-time jobs, casual and intermittent jobs, dead-end jobs, menial jobs, and degrading jobs. The inability of the statistics to reflect the quality of employment has in a sense limited their usefulness as a gauge of public policy. All this has had its effect—and, other things being equal, it has tended to make a given level of unemployment less tolerable than heretofore.

The acceptance or nonacceptance of given patterns of growth change has also undergone change. Up to 1960, we accepted the business cycle, with rates of growth ranging from mildly negative to as high as 8 percent and 9 percent a year. Now, with the increased usage of powerful fiscal and monetary tools, this is no longer deemed acceptable. The aim today is to keep growth on an even keel, with the acceptable range in any given year between a low of 3 percent and a high of about 5 percent. Unfortunately, this goal isn't always met. Even so, it does signify a change in what we would like to see happen.

Percentage changes in prices also take on different connotations over an extended period of time. Remember the 1950s, when a 1 percent to 2 percent annual rate of inflation was deemed "disturbing." Not so today, when 3 percent or even 4 percent advances are regarded as almost inevitable. Here again the reason is essentially the same: changing ground rules. Specifically, the increasing accent on ecology, safety, full employment, etc., has made it a lot more expensive to produce. Ergo, to expect prices to follow the pattern of one and two decades ago is clearly unrealistic.

Indeed, some economists argue that to force price rises into the 1 percent and 2 percent mold of the 1950s could be disastrous. Just such an attempt was made in 1969 and 1970—and it plunged the American economy into its first full-fledged recession in nearly 10 years.

FOUR

MORE ABOUT INTEREST RATES

Skeptics who still think that the flap about "truth in lending" is a tempest in a teapot would do well to consider the results of a recent Federal Reserve Board interest rate study.

This conservative-minded group of banking analysts (and one that would hardly be expected to be biased in favor of consumers) found that as late as 1972 consumer finance companies were charging an average annual interest rate of 21 percent — a figure that would make even the most penurious usurer blush.

Hard to believe? Sure. But nevertheless true.

What made this figure so astounding was that businessmen at the very same time were paying as little as 5½ percent (the so-called prime rate) for their money. In other words, John Q. Public, who historically has been only a slightly greater credit risk than his business counterpart, was being asked to pay four times as much for the privilege of borrowing money.

What to do? Taking the bull by the horns a few years ago, the Congress of the United States passed a comprehensive truth-in-lending law. In essence, it called for a halt to the practice of keeping the effective rate of interest hidden from the borrower. More specifically, it provided that every borrower must be told how much it costs him—in terms of a true, or effective, annual interest rate—to borrow money. In a sense, it demanded that lenders do for consumers what they have always done for businessmen—that is, state explicitly the annual rate of interest being charged.

For some lenders this created quite an upheaval. Take the typical charge account that most middle-class consumers maintain at their favorite department store. Prior to passage of the truth-in-lending law, the charge was invariably reported as, say, 1½ percent per month on the unpaid balance. Not so any more. Each and every such charge must be stated in terms of an equivalent annual rate—18 percent in the case of the example cited immediately above.

Department stores fought the disclosure measure on the grounds that most users of revolving credits don't let their bills remain unpaid for such a long period of time. But this was begging the point. Even if you pay only one month's interest charge, knowledge that cheaper rates exist might well lead to borrowing the money more cheaply elsewhere (perhaps at a bank)—and then using the proceeds to pay the charge account before interest began to accrue.

At the time the law was passed, another question arose: How do you calculate the true interest rate? After much hemming and hawing and statements to the effect that it couldn't be done, lawmakers settled on the so-called actuarial method. In simplest form this involves a redistribution of the dollar finance charge in accordance with a series of unpaid balances under an installment contract. It is similar to the way mortgage interest rates are calculated. The net effect was to outlaw the use of other (and in most cases deceitful) methods of computing interest rates.

The two types that were eliminated once and for all were the "add-on" and "discount" methods of computation. Both had one thing in common. They led the unsophisticated borrower into believing he was paying a lot less for the use of money than he really was.

CALCULATING THE TRUE INTEREST RATE

A few simple examples can translate these deceptive approaches into effective actuarial rates—and can perhaps best define the magnitude of the deception.

Let's start with a borrower who wants to borrow $100 for a year. Normally a bank would offer the potential customer the "add-on" option. Under this technique, and assuming a 6 percent interest rate, (1) the $100 would be received by the borrower and (2) the loan would then be repaid in 12 equal monthly install-ments that added up to $106 over the year. It was this $106 that led the bank to claim only a 6 percent interest rate. The borrower was being given $100 and was expected to pay back only $106.

But a few moments' thought should reveal the fallacy of this claim. Specifically, the borrower has the effective use, on the average, of only about half the $100 over the 12-month period (remember that he is paying back part of the principal with each monthly installment). Indeed, calculating on the more realistic actuarial basis would show that the true, or effective, annual rate under this arrangement comes to 10.9 percent—nearly double the bank's apparently low rate.

It should be pointed out that the bank can still use this add-on method. But now it must also give the borrower the equivalent of the actual effective rate, too. Thus today a bank continues to employ its old approach—but must tell you that the effective rate is 10.9 percent rather than the "6 percent" that it was employing before the truth-in-lending law went into effect.

Nor is 10.9 percent the highest effective rate possible when a 6 percent nominal rate is quoted. Here is where the "discount" approach enters the picture. The payback is still the same as in the add-on method—only this time the bank takes off its 6 percent before the loan is actually made. Specifically, you, the borrower, get only $94—and are then expected to pay back the $100 principal in 12 equal installments.

As in the case of the add-on method, John Q. Public has the effective use, on the average, of only half the borrowed amount over the year. But there's a difference. This time he is paying $6 for the use of $94—where before he was paying the same $6 for receipt of a bigger $100. It is not surprising, then, that the discount method carries a somewhat higher effective interest charge—11.5 percent—compared with the 10.9 percent under the add-on method.

So far the discussion has centered on techniques used by banks. But these rates—as high as they are—are low when compared with those asked by department stores and finance companies. As such, banks in many cases may still be your best bet for consumer-type loans. But it should be pointed out that there are still two limited sources for even cheaper borrowing that deserve honorable mention. First, there is the passbook loan, which involves borrowing against money already on deposit in a savings account. Then there are life insurance loans—that is, borrowing against your life insurance policy. Such life insurance loans can often be obtained at a straight 5 percent or 6 percent interest rate—far under the prevailing market cost of money.

WHAT TRUTH IN LENDING DOES
AND DOESN'T DO

It should also be noted in passing that some creditors often levy a service charge or carrying charge—or some other charge—instead of interest. Or perhaps they may add these charges onto the regular interest fee. Under the truth-in-lending law, they must now total all such charges—including interest—and call the sum the "finance charge." And then they must list the annual percentage rates of the total charge for credit.

On the other hand, the truth-in-lending law does not fix interest rates or other credit charges. (Your state, however, might have specified upper limits. If so, such limits would then apply to residents of that state.)

Chances are, then, that if the monthly statements and other materials you receive from companies giving you credit look different now, it doesn't necessarily mean they've changed their rates. Probably all they've done is to change their way of showing them in order to comply with the new law.

Your department store bills, for example, should now list both a monthly rate (say, 1 percent, 1½ or some other number) and the annual percentage rate. This may be 12 percent, 18 percent, or some other percentage. But it will usually be 12 times the monthly rate.

The law also provides guidelines for the advertising of credit terms. It says that if a business is going to mention one feature of credit in its advertising, e.g., the down payment, it must mention all other important terms, such as the number, the amount, and the period of payments that follow. If an advertisement states, "Only $2 down," it must also state, for example, that you will have to pay $10 a week for the next two years—if that, indeed, is the arrangement. Here again the intent is to provide the consumer with full information so that he can make informed decisions.

One important by-product of the truth-in-lending law: it has made potential borrowers stop and think before plunging ahead. The appeal of the old "dollar down and 52 weeks to pay" argument isn't nearly as enticing when the true interest rate is brought to the fore. And if this doesn't stop the potential borrower, it at least alerts him to other borrowing options—or perhaps it might induce him to make a larger down payment or maybe even pay cash to avoid the sky-high interest charge.

Before leaving this section, it should also be pointed out that lenders "went down fighting"—with many trying to circumvent the law for years after it was originally passed.

"In compliance with the Truth in Lending Act, I'm required to tell you that if you miss over two payments you'll wind up in the hospital." (© *The Wall Street Journal.*)

It was only in 1972 (some three years after it was passed) that the law caught up with several large department stores that were charging too much interest on revolving credit plans. Up till then these big sellers had neglected to tell buyers that they were being charged interest on the previous month's balance rather than the average daily balance of the current month. When faced with the dictum of (1) either telling customers they were paying a finance charge on bills already paid or (2) switching to an average-daily-balance approach, the stores had little alternative but to choose the latter. Ergo, another victory for the consumer.

THE HIGH COST OF MORTGAGES

Potential house buyers are often duped or misled by relatively low interest rates (10 percent or under) on mortgage loans. Not everyone realizes that these small rates add up to large dollar costs, both because of (1) the large sums of money involved and (2) the extended period of the loan.

Figure 4-1 points up just how much such mortgage payments can add up to. Thus even if you are lucky enough to get, say, a 25-year, 8 percent mortgage, by the time you pay it off you would have given the bank $2.30 for every dollar borrowed.

Moreover, since most mortgages are for bigger amounts, longer payment periods—and sometimes higher interest rates—chances are your own mortgage calls for even higher bank paybacks.

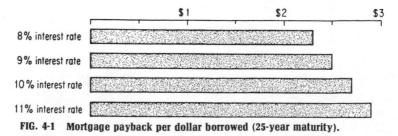

FIG. 4-1   Mortgage payback per dollar borrowed (25-year maturity).

Actually, there is no way out of the box—except that one should always attempt to keep the factors that increase payments (maturity, loan amount, and interest rate) down. The first, of course, depends upon how big a monthly carrying charge you can afford, the second on the down payment you can scrape up, and the last on the state of the money market.

You can probably do something about the first two factors— if you try hard enough. And the payoff is certainly worth the effort. Thus should you reduce a mortgage from 30 years to 20 years, you could probably cut as much as 50 percent off your interest charges.

On the other hand, there is little control over the interest rate— unless, of course, you can persuade the lender to insert a clause which gives you the privilege of prepaying the mortgage in the event of a cash windfall. Sometimes this is worthwhile even if the bank insists upon a small penalty charge.

COMPOUNDING THE FELONY

The mortgage costs alluded to above point up how a small difference in rates can result in a tremendous payment difference over the longer pull. In the case of mortgages, this gap is primarily due to the long period of time involved. But in many other instances, the sharp difference is due to the compounding process (applying a growth or interest rate not only to the original figure but also to all the increments of preceding periods). The difference between compounding and not compounding can be significant. Using simple interest (rather than compound interest) at an annual rate of 10 percent would double your money in ten years. But with compounding, the doubling would occur in only seven years.

There is, incidentally, a rule of thumb for calculating the compounding effect. Simply divide the interest rate into the number 70. The result is the number of years it would take for your money to double. In the above example, the 10 percent rate divided into 70 yields the 7 years needed to double your money.

But compounding affects more than just interest rates. Growth rates are equally influenced. An economy that grows at 2 percent a year will be a lot worse off than an economy that grows at a 4 percent rate after a period of years because of this compounding effect. Again let's use our "70" rule of thumb. The 2 percent-a-year economy will double its output in 35 years. The 4 percent-a-year economy will double its output in half that time, or 17½ years. Thus just by accelerating the economy an additional 2 percent a year, you can achieve a doubling of your standard of living in roughly one generation instead of two.

Similarly, rates of inflation are also affected by compounding. Substitute the word "price" for "growth" in the above example, and your purchasing power will be halved in either 17½ or 35 years—depending upon whether the annual rate of price rise is 4 percent or 2 percent. Again, small differences in one year add up to big differences where several years are involved.

If you still doubt this fact, consider the difference between a 1 percent rate of inflation and a 5 percent. At 1 percent, the purchasing power would be halved in 70 years. At 5 percent, the halving (or doubling of the price level) will take place in only 14 years. Note here, too, how the "70" rule applies: 5 percent divided by 70 years equals 14 years—and 1 percent divided by 70 years equals 70 years.

CALCULATING THE TRUE RATE OF RETURN ON INVESTMENTS

Compounding and effective rates are also of vital concern in making any investment decision. The average investor is usually faced with a myriad of investment opportunities—and it is crucial that he be able to calculate the true rate of return in order to maximize his profits. The problem, however, is that the computation of the effective rate is quite complex.

To avoid problems, some take the easy way out—and only want to know how long it would take them to get their invested money back. Known as the "payback" period, this is quite simple to compute. Thus, if you decide to buy a $10,000 parcel of real estate and the additional income after taxes is $4,000 a year, the payback is 10,000/4,000, or 2½ years. The problem here is that this measure ignores the years after the payback period. In other words, liquidity rather than profitability is stressed.

On the other hand, the payback period does have some advantages. First, you know when you will be getting your money back—a pretty important piece of information in these days of high working capital requirements. The payback period can also be regarded as a valuable gauge of risk. A quick return means that your money will be coming back in the near future, when revenue predictions have the highest degree of reliability and validity.

A better measure of profitability is the rate of return (profit over investment). Here, again, there is an easy way and a complex way—with the more complicated approach the more meaningful one.

The simpler approach—known as the "accounting" method—expresses average annual net income as a percent of capital outlay. The trouble is that this approach ignores the fact that a dollar earned today is worth more than a dollar earned well out in the future. More specifically, at a 6 percent interest rate, the promise of a dollar today is worth twice as much as the same dollar promised some 10 years from now.

This "timing" shortcoming is eliminated by what has come to be known as the "discounted cash flow" method, which takes into account when all earnings are received.

A simple example can best illustrate the difference between the accounting and the discounted cash flow methods. Again assume the potential purchase of a $10,000 parcel of real estate. But this time you expect net cash inflows in the form of rent to amount to $1,000 a year over the next decade. Once again, the basic question is the rate of return.

Under the accounting method, where you give equal weight to all payments, you would, of course, wind up with a $1,000 average income over a $10,000 investment—or a 10 percent rate of return. But under the discounted cash flow approach, payments over the latter years are worth considerably less than $1,000. While the mathematics is a bit complex (though somewhat akin to the compounding effect), it can be shown that the rate of return under this method will be considerably less than under the accounting approach—something in the order of only 7¾ percent a year.

The fact that the accounting method overestimated the true or real rate of return is not surprising—since the far-out and hence less valuable revenues were given the same weight as near-term ones.

In any case, the time and effort needed to get the discounted cash flow answer is well worthwhile. For example, if a bank were offering an 8 percent interest rate, you might have been tempted to invest in the real estate deal if your calculation were based on the 10 percent accounting rate of return. But in truth you would have been making a mistake, for the true or effective rate of return would have been only 7¾ percent, and you could have bettered this with absolutely no risk by simply depositing your money in the bank.

FIVE

EQUATING SHADOW WITH SUBSTANCE

Not everything can be forced into a quantitative straitjacket. Common sense would suggest as much. Yet more often than not this obvious fact is forgotten, and numbers are dreamed up— either to prove a point, to justify an action, or, sometimes, just to provide an illusion of reality.

In a sense, this is the most dangerous type of distortion, for it often dupes the perpetrator as well as the recipient. The most obvious example in recent years involved the measurement of progress in the Vietnam war by changes in the "kill" count. It lulled many a Pentagon general into a false sense of security when no such feeling was warranted. But the pitfall is one that almost all of us fall into some time or other in our lives: the relating of an easily measured statistic (e.g., kill ratio) with a basically complex problem (e.g., the war in Vietnam).

Part of the problem lies in our cultural worship of numbers. Open up your newspaper and look at the unemployment ratio, the latest GNP trend, the Dow Jones average, or even baseball batting averages, and one fact becomes clear: we have become a nation of scorekeepers—even in cases where such scorekeeping is neither desirable nor even warranted.

It also matters not if the score is even pertinent. Attach a number to anything, and the subject suddenly takes on the aura of deep significance. Sometimes this quantification borders on the ludicrous. The first man in history to strike out four times in the third game of a world series makes the headline on the sports page—even though the news is clearly meaningless as far as the ball player or even the ball game is concerned.

The history of the past decade is replete with other such examples—many of them harmless, but many others leading to serious strategy errors on the part of both individuals and governments. For no matter how ingenious or brilliant one is, he cannot measure the unmeasurable or quantify the unquantifiable.

The sad part of all this is that men of integrity, brains, and diligence have fallen into this trap. In many such cases, self-deception rather than dishonesty is to blame.

A MANY-FACETED TRAP

The advertising man who measures a TV program's rating and then equates this with the program's ability to sell a product provides perhaps the classical example of the shadow-for-substance pitfall. One doesn't have to be particularly astute to know that a one-to-one relationship just doesn't exist here. It is quite conceivable, for example, that the vast majority of people who watch a popular program are in no way influenced by the accompanying commercial. On the other hand, it is possible that a program with a relatively poor rating may have great success in influencing its few listeners to buy brand A rather than brand B.

Nevertheless, the advertising profession continues to opt for this questionable approach. And the reason isn't too hard to find. It's relatively easy to quantify the number of people who watch a given program, but quite difficult—indeed, virtually impossible—to determine the number of listeners who will ultimately go out and buy the product in question. In short, the advertising man is taking the easy way out. The danger here, of course, is that the easy way out may well be the wrong way out.

At other times the danger comes from attempts to assign numerical values to areas where no such quantitative assessment is possible. A good example: the recent attempts to develop indicators of social conditions. Thus statisticians are now actively seeking ways of measuring the "well-being" of society.

The problems involved here are virtually insoluble. "Well-being," for example, certainly depends on the volume of useful goods produced. But who is to define what is useful and what is wasteful? War goods are often put into the latter category. But can the same be said of annual auto model changes that give some buyers a feeling of satisfaction? Then, too, do higher income and lower employment totals necessarily add up to contentment and security? In short, is happiness an ever-rising GNP?

The answer to this last question is obviously, No. But if esoteric concepts such as happiness are to be included in our "well-being" yardstick, how does one go about measuring them?

There are two kinds of problems that are basically involved here: one mechanical and the other philosophical. Take the mechanical difficulties first. It is clear that sales of autos and eggs can be combined to make GNP because both have prices stated in dollars. The price is the same for everybody. But combining social values isn't that simple. The elements that make up a yardstick on crime, for example, are seldom additive. Thus there is no way of coming up with the answers to such questions as: How many robberies equal a murder?

Another mechanical difficulty: How much weight should be given to the physical aspects of a welfare index and how much to the social aspects? Is love worth more or less than money to the average human being, and, equally important, how much more or less?

The philosophical roadblocks are even more foreboding. No one would deny, for example, that a sense of well-being is subjective. Under a given set of living conditions, Mr. A may feel quite content, Mr. B tolerably well, and Mr. C miserable. Yet all are faced with the same objective reality. Which view, then, are we to believe when formulating our index of well-being?

Another philosophical poser: How does one quantify factors which defy our normal concepts of numbers? If someone gives $10 to another person, he is obviously $10 poorer. On the other hand, if he provides information to another, he doesn't gain or lose. And if he gives love, he himself has more love. The point is this: with noneconomic factors, 2 plus 2 doesn't always equal 4. They make 5, 6, and sometimes even 7.

Ergo, a good many statisticians have thrown in the towel. Their conclusion: a single generally acceptable index of welfare cannot be constructed.

Nevertheless, the search for the Holy Grail goes on. The *Economist* magazine of London took a stab at it not so long ago. This renowned publication compiled an index for 14 countries based on what it considered 15 important social indicators—including such things as car ownership, the divorce percentage, economic growth, and the ratio of TV sets to people.

The trouble was that, despite the inclusion of a few socially oriented yardsticks, the index ignored many other noneconomic factors which are considered important to the quality of life. As such, the *Economist* index proved to be a dud.

Similar types of problems spill over into the business area. Thus corporate executives and their critics increasingly agree that the concepts of growth and profit, as measured by traditional balance sheets and profit and loss statements, are too narrow to reflect what many modern corporations are trying to do. But they're still at a loss on how to deal with the problem.

How, for example, can a profit and loss statement be made to reflect the good a company does by assigning some of its personnel to advising minority businessmen struggling to succeed in the ghetto? How can the installation of pollution-control devices at a factory be shown on the bottom line as a positive accomplishment rather than as a drag on productivity? How can the expense of hiring dropouts and putting them through company-financed training programs be reflected on the credit rather than the deficit side of the ledger?

Despite these formidable problems, some halting attempts are being made to work up what are becoming known as "social audits." A few firms are also toying with what could be called "performance audits," trying to measure the progress of corporate programs against well-defined standards. Such attempts so far have been made in the areas of hiring and pollution—sectors where local or national guidelines have been spelled out.

Unfortunately, such guidelines are few and far between. Indeed, in many sectors of social concern, there are no accepted standards against which to measure performance. Moreover, even where standards do exist, social critics are constantly pressuring companies to do more—for social responsibility is a moving target. In short, the standards never stand still long enough to provide meaningful yardsticks.

In any case, it would seem a great mistake for companies to try to put exact figures on values that are still unmeasurable by any known technique. Few today would deny there is a deceptive precision about figures that can imply certainty where only the crudest approximations are possible. Thus, any dollar figures on a program to employ marginal workers will be misleading because the out-of-pocket costs can be measured fairly accurately but the benefits can only be guessed.

The day may come when it is possible to present a social income statement and a social balance sheet stating costs and benefits in dollars. But for the present, such a report would conceal more than it revealed. It has taken financial accounting over 500 years since the invention of double-entry bookkeeping to reach its present, still-imperfect shape. Social accounting may move faster, but it still has a long way to go.

But not everybody agrees with this timetable. Thus Ralph Nader, the consumer advocate, feels such social quantification can be accomplished quickly. What we need, he says, is more stringent laws and regulations to hold companies accountable for social costs they now are able to avoid. "The problem is being overly complicated," he says. Nader suggests changes in the cost-benefit ratios, with perhaps the levying of fines running into the millions for firms that fail to meet specified standards. This controversial consumer advocate would even go as far as to send some corporate executives to jail. "You'd be surprised how fast they could quantify that."

OTHER PITFALLS OF SOCIAL MEASUREMENT

But the above comment is too much of an oversimplification. Aside from deriving measures in the social sphere, there's the problem of discovering neat statistical relationships among the social indicators themselves, as well as between the social and economic indicators.

The nice thing about a business indicator—say, a price or production index—is that it is part of a statistically related economic system. It is more or less understood that there is a bag of tricks that can be brought to bear to affect these yardsticks. For example, reduce labor costs or demand, and sooner or later prices are bound to fall.

On the other hand, things are not that neat in the social sphere. Spending more money on education, for example, does not necessarily result in higher educational attainment. Many educators are now convinced that no school, no matter how well run, can turn the majority of ghetto children into college graduates. They rightfully point out that cultural influences, independent of school experience, play the crucial role in education and professional achievement.

To be sure, other things being equal, a good education is better than a bad one. But unfortunately, this "other things being equal" condition—statisticians like to refer to it as "ceteris paribus"—rarely exists in the real world.

The approach of pouring more and more money into crime prevention has led to similar results. Again the one-to-one relationship between the two variables in question (money and crime prevention) simply doesn't exist. Crime is only partially attributable to the lack of police protection. Much of it stems from deep-rooted cultural disturbances which cannot be treated by opening up the spending spigot.

One doesn't have to look very far to back up this thesis. Crime has continued to rise despite the pouring of hundreds and millions of additional dollars into police protection. Indeed, sometimes such attempts have a curious reverse effect: reports of crime actually go up—not so much because crimes have increased, but because there are more law enforcement people around to pick up the crimes.

At other times, crime indexes can decline because they can easily be manipulated by politicians. If you are up for reelection, for example, you may well decide to twist the arm of your police commissioner to make sure that less crime gets reported on the police blotter during the crucial few months before election.

At other times crime is unreported because of victim reluctance to bring it out into the open. Thus, figures compiled during recent intensive interviews in 5,000 homes and 1,200 commercial establishments revealed that the number of rapes and robberies actually experienced ran twice as high as reported. More significant, the number of aggravated assaults reported to the interviewers ran at five times the rate recorded by the local police.

The reasons given for not reporting these crimes included involvement of friends and relatives in the crime, a belief that the crime was too petty for police action, and fear of retribution for calling the police.

Overall, the study's findings support the argument of critics of the *Uniform Crime Reports,* published by the Federal Bureau of Investigation, that now serves as the official measure of crime in the nation. It lends credence to their complaint that reports which are compiled by local police agencies are a poor measure of crime in the country.

Intelligence tests are another area open to serious question, for they too attempt to measure the unmeasurable. The fact is, no matter how much scientific jargon you want to attach to such a test, it is essentially nothing more than a mechanical instrument attempting to measure the nebulous concept we call intelligence. To be sure, scores on a test can be calculated rather easily. But what does all this tell us? Probably very little, for the property being measured (intelligence) remains little more than an abstraction. If nothing else, educators must take more care in drawing inferences from the mechanical to apply to the abstract. There may be some relationship, but it is not the simple one suggested by proponents of the testing approach.

If there is still any doubt on this score, a recent experiment involving the so-called intelligence levels of minority groups should be the convincer. Tests were given in English—and for the first time in the groups' native languages. The results: scores were much higher in the latter group. In other words, a change in what heretofore had been regarded as a near-perfect yardstick of intelligence resulted in major differences in the final results.

Another problem: All too often, educators have a vested interest in maintaining the myth of the infallibility of these tests. If they admit to weakness in the tests, they are undermining the entire justification for their function.

Equally difficult dilemmas exist in measuring some of the more common forms of business and economics data. In a very real sense inflation, poverty, and the hard-core unemployed are little more than man-made, socially defined concepts. Statistics in these areas are therefore on just as shaky ground as those noted above for education. Again, in all too many cases, the numbers are essentially only partial statistical truths rather than objective realities.

Much of this stems from self-deception. By taking the easy way out and oversimplifying relationships, these social scientists have convinced themselves they have the answers to evasive, many-sided problems. The basic difficulty, of course, is that they are just seeing the top of the iceberg. The 90 percent of the problem which is submerged is usually far more important than any small amount of information revealed by the statistics.

MEASURING POVERTY

Part of the problem here lies in the fact that margins for misjudgment are often ignored by the people who play this "numbers" game. Attracted by the appearance of objectivity and precision, they keep their eye squarely on charts and tables which may be incomplete, obsolete, or even irrelevant. Eventually, many come to believe that poverty really is a condition of having less than a given annual income, that economic nirvana really is a situation where the national jobless rate has dropped to 4 percent or less, and that 4.2 people living in 2.7 rooms is really a significant cutoff point between adequate and inadequate housing.

For those still unconvinced, a look at the failures of the 1960s may be in order. At that time the poverty level was defined at a given dollar income. Great joy was expressed by statisticians over subsequent years when it appeared that more and more people were climbing over the "poverty line."

But a curious thing began to happen. The higher up on the statistical income ladder these people climbed, the more dissatisfied, disenchanted, and militant they became.

A paradox? Not really. The problem was that a simplistic cut-off figure was inadequate for separating the poor from the non-poor. The truth of the matter was that many of these people who had jumped over this statistical dividing line still felt poverty-stricken. Again, a mechanical measure (dollar income) proved to be woefully inadequate as a measure of a sociological abstraction (poverty).

To be sure, nonabstract factors also contributed to the failure of the antipoverty program. Specifically, while the income of the poor was rising, the income of the middle class was rising even faster. This, in turn, widened the gap between the two classes — thereby breeding even more discontent among the aspiring poor.

THE GNP DILEMMA

Even purely economic yardsticks are coming under critical fire these days. Consider GNP (the market value of all goods and services produced). Many claim that GNP growth deceives us into complacency. While statistically everything seems to be getting better, environmentally everything seems to be deteriorating.

Aside from moral issues, much of the problem stems from the fact that you get nothing for nothing. Each increase in material standards has seemingly been achieved by a decrease in environmental standards. This has prompted some to propose that GNP be adjusted to take account of the depreciation in our environment. When there is a proper recording of the minuses as well as the pluses, we may well discover that the GNP, which has been deceiving us all along, is a good deal lower than we think it is.

How much does all this environmental depreciation amount to? While it's hard to put a dollar value on breathing clean air and swimming in clean water, some people feel that as much as one-third to one-half of our annual GNP advance has been illusory. Put another way, GNP tells us only of quantity and nothing of quality, neglecting to make any subtraction for the losses in amenity which are the consequences of overcrowding, pollution, transportation failures, housing shortages, and the sheer pressure people exert upon one another.

There have been other, nonenvironmental attacks on the GNP concept, too. Does military production, a major part of our GNP, really add to our standard of living? And do the billions of dollars paid for repairing shoddy merchandise (also counted in GNP) really represent an advance in our standard of living?

Then, too, virtually any type of disaster, personal or national, will cause the GNP to rise rather than fall, according to F. Thomas Juster of the National Bureau of Economic Research. Thus if a man's wife should die and he is forced to hire a maid, the GNP goes up, because only time spent on a paid activity finds its way into the statistics. Yet, of course, the same work is now being done by the maid as had been previously done without compensation by the wife.

Other GNP shortcomings: If workers gain more leisure (certainly a plus as far as the workers are concerned), GNP does not go up. Nor is this all-inclusive measure reduced by the time wasted in traffic jams and other negative pursuits.

Mr. Juster adds that no account is taken of "the unwanted side effects of economic activity—rivers that cannot be used for recreation or parks cluttered with disposable bottles." If air pollution makes it necessary to repaint a house, the job actually boosts GNP unless the home owner does it himself. And even then, the cost of the paint would show up in GNP totals.

Even government officials admit to a problem. In a section of the 1971 Economic Report of the President dealing with national priorities and national output, it is noted that:

> While the Nation has been engaged in a new and earnest soul searching about the role of growing material affluence in the good life, it is probably true that in general the American people prefer a rapid growth of GNP and its consequences.

However, an important qualification to this statement appears shortly thereafter:

> This is not to say that growth of measured GNP is an absolute to be furthered at all costs. As individuals and as citizens we clearly do many things that reduce the growth of GNP, and we fail to do many things that would accelerate it. This is perfectly reasonable; growth of GNP has its "social costs" and beyond some point they are not worth paying.
>
> Man wants more than is counted in GNP. People's values change. Conditions of life change. These may lower the point beyond which more growth of GNP is not worth its costs.

To sum up, after a decade spent in the spectacularly successful pursuit of a rising GNP, the United States has suddenly begun to have second thoughts about growth as a goal for national policy. Up till recently the exceptional growth rates of the sixties were just about taken for granted as a model for the seventies. Yet more and more people are beginning to see economic growth itself as the basic cause of the environmental deterioration that has become one of the nation's most pressing problems.

In a sense, as the carpet of increased choice is being unrolled before us by the foot, it is simultaneously being rolled up behind us by the yard. It may well be that modern Western civilization has reached the point of diminishing returns, where the negative environmental factors generated by increased production outweigh the increase in the number of cars that rush us from place to place or the increase in the number of transistor radios that make a day on the beach a hell on earth.

Is GNP retrenchment the answer, then? Maybe. But there are other possibilities, too. These might include a revision of the GNP concept to include environmental improvement as part of national output—that is, we might include an improved environment as a dollar-evaluated plus. Another solution would subtract out the dollar value of environmental deterioration—on the assumption that if we see the downward pressures, we may be more willing to do something about them. Still another suggested solution: penalize polluters, primarily by taxation. This would raise the cost of production of polluters, and hence either discourage their output or persuade them to take corrective antipollution action.

THE ROLE OF THE COMPUTER

In a sense, the computer has served to heighten the tendency to measure the unmeasurable. Too often one tends to forget that these electronic monsters, since they work only with quantitative data, often overlook equally important qualitative aspects of a problem.

In 1971, for example, all the statistical indicators—income, wealth, consumer spending surveys, and the aging stock of consumer durables—pointed to a substantial rise in retail sales. Computer models were almost unanimous in predicting close to a 15 percent rise over recession-ridden 1970 levels. Yet when the dust finally settled, the year-to-year advance came to a disappointing 7 percent—a figure that dwindles to near zero when price rises are taken into account.

Why the goof? Hindsight suggests that the computer was just not able to evaluate such subjective factors as social unrest, the general feeling of malaise that prevailed at that time, and fears of unemployment. The truth is that these latter factors are perhaps even more important than such easily quantifiable statistical yardsticks as purchasing power when it comes to making the complex buy or no-buy decision—at least in the short run.

Sometimes ignoring these nonquantitative yardsticks can result in considerable financial loss. Thus a few years back the government decided to buy rather than rent computers—with the ultimate rent-or-buy decision itself based on computer analysis. But this proved woefully inadequate. To be sure, the computer could quantify the dollars-and-cents savings involved by going the "buy" route. On the other hand, it was unable to quantify what was then shaping up as a crucial obsolescence factor. In any event, based on the computer recommendation, Uncle Sam bought its computers—only to find the computer industry coming out a year later with a new generation of computers that made its multi-million-dollar purchase a financial fiasco.

Summing up, there's room in the world for the quantifiable and the nonquantifiable. And to stress only the former is fraught with danger. Indeed, all the basic distortions discussed in this chapter revolve around the tendency to use numbers, even when such numbers are spurious or of questionable validity or, where valid, are not crucial to the problem at hand.

Nor are the mistakes the province of any one group of individuals. The same liberals who have little trouble seeing through the shortcomings of the GNP concept are firmly convinced that more money on education will solve the nation's minority problems. Similarly, archconservatives who rightfully point out that no such correlation exists between educational outlays and the quality of education shell out endless sums of money for police protection in their elusive pursuit of "law and order."

Such inconsistencies—and there are literally thousands of them if one takes the time to ferret them out—can be overcome only by taking a more critical attitude toward all quantitative data. The basic question to ask is not whether the numbers being presented are correct or accurate in the narrow sense of these words, but rather whether they are relevant. No matter how you slice it, accuracy without relevancy is meaningless.

SIX

SPOTLIGHTING THE SPURIOUS

Some years ago, a Midwesterner who had a penchant for numbers remarked that the Mississippi River was exactly 1 million and eight years old. How could he be so sure? Well, it turns out that just eight years previously he had heard the state geologist say that the "Old Muddy" was a million years old.

The anecdote is humorous because it's so obviously ludicrous; yet literally thousands of times a day, people—many of whom should know better—are using numbers and figures which are irrelevant, meaningless, or sometimes even both.

In many cases, of course, this spuriousness is intentional. Indeed, a goodly portion of our advertising profession today is devoted to dreaming up important-sounding statistics that have little relevancy to the message being transmitted.

Percentages are in many ways tailor-made for this type of manipulation. What does the advertiser mean when he says his brand A is 20 percent cheaper than competing brand B? The initial outlay may be cheaper, but other legitimate costs such as operating expenses and depreciation may actually make it more worthwhile to buy brand B (more about this below).

Indeed, the entire area of percentages has been so rife with distortions and misconceptions that two separate chapters (Chapters 3 and 4) have already been devoted to this subject.

But percentages are by no means the only problem areas where spuriousness is involved. This is a pitfall that affects the entire range of numbers and figures and also permeates our own thinking as well as that of the advertiser, with his obvious ax to grind.

Thus the whole question of prejudice is laced with irrelevant statistical relationships. How many times have you heard people say that because the crime rate is highest among blacks, this is prima facie evidence that blacks are more prone to crime? But the real truth (if reputable sociologists are to be believed) is that it is the condition in which many blacks live that breeds crime, and that whites living under the same conditions might act similarly.

This is all tied up with cause-and-effect misconceptions — and the tendency on the part of many to think in terms of cause and effect when no such connection is warranted. A rundown on this and some of the other more popular attempts at spurious statistical presentation follows.

MEANINGLESS ACCURACY

In far too many instances the statistical charlatan will attempt to leave the reader or listener with the impression that his facts and figures are far more accurate and meaningful than they really are. One of his favorite techniques: excess precision. Somehow a rounded-off number never sounds quite as impressive as one carried to three or four decimal places—even though such greater accuracy may well be meaningless or not of any more use to the recipient than a rounded-off figure.

This is often the favorite tack of samplers. Tell someone that 71.3 percent of the people prefer brand A to brand B, and he will be a lot more impressed than, say, by a statement indicating a 70 to 30 preference break. If he would stop to think of all the possible errors involved in the sampling process, he might realize that the 71.3 percent figure is meaningless as far as providing more information than the rounded-off figure of 70 percent.

But even those with no ax to grind are sometimes guilty of this kind of spurious accuracy. The high school or college student asked to divide one number into another will carry his answer to three or four places beyond the necessary decimal place—somehow always believing that one additional place will yield a small additional amount of accuracy. Similarly, the retailer asked to calculate average sales per customer may come up with a figure of, say, $10.48, while a more rounded-out estimate of around $10.50 would have been sufficient for his planning of future inventory, advertising, and promotional strategies.

There's a purely statistical aspect to accuracy, too. Mathematicians tell us you get no more accuracy out of an answer than the accuracy of the data you put in. To say this in simple terms, if data input is rounded off, then so should data output be rounded off. In any event, watch out for significant digits, limiting them to no more than those contained in any one of the figures involved in the calculation. Thus if one bit of input has three significant digits and the other four, then your answer should contain no more than three significant digits. Anything more detailed would be statistically meaningless.

UNNECESSARY DETAIL

Too much detail can be as self-defeating as too little, weighing down decision makers in a morass of insignificant minutiae as well as adding to costs. Thus if an auto firm is thinking about allocating funds for capital expansion five years out into the future, it makes little sense to work up model-by-model forecasts. That's because the overall unit sales estimates will generally be sufficient to signify how much additional plant expansion is necessary. In short, the model breakdown adds nothing toward the solution of the problem at hand.

Turning to an example in the consumer area: A voter, in determining whether or not a certain party has done a good job in containing inflation, doesn't have to know the price history of 200 different commodity groups. Give him the overall rise in the cost of services, food, and consumer durables, and he has enough information to reach an intelligent decision. The figures showing that beef prices are up and that they were partially offset by declines in pork add little relevant information.

ADVERTISING DISTORTIONS

In a sense, advertising has given those who would use spurious ploys an unequalled opportunity for pulling the wool over the unsuspecting consumers' eyes. Whether it be TV, radio, newspapers, or whatever, the numerical statements made about products often border on the ludicrous. And when they don't, so much of the pertinent information is left out as to make the advertising message deceptive or even downright misleading.

On the latter score, probably nothing has been quite as costly to buyers in recent years as the false advertising involved in the selling of expensive condominiums. The seller is right there ready to give you the base price, but in almost all instances he will conveniently forget to tell you about all the extra services for which big additional fees will be required. In some instances buyers report paying as much as 25 percent and 30 percent more in monthly payments than the original brochure had promised.

Less than full disclosure has also angered those who have

been buying from catalog showrooms. This is a retailing technique whereby consumers can make their selection at home (from a catalog), come into the store, examine the display sample write up their own order, wait while the item is removed from a back room inventory, and then finally carry home the merchandise themselves.

All this is supposedly done to permit bargain prices. And sometimes it does, but not nearly as often as some customers had been led to believe. One bone of contention: misleading price advertising. Specifically, the typical showroom catalog lists two prices: the so-called manufacturer's list or suggested retail price, and the special price the customer is supposed to pay.

It is the difference between these two prices that bugs the buyers. To be sure, most catalogs carry a disclaimer, saying something like "Manufacturer's list price does not necessarily represent the usual price at which these items are sold at retail." But many customers do not see or pay attention to this notice, and they think that they are saving the difference between the quoted discount and list price. The point is that if most other stores sell the product for less than the printed list, then the catalog which suggests a list less discounted price saving is quite misleading.

Even more insidious are the downright untruthful ads. These can fall into several different categories. The most obvious, of course, is the advertisement that quotes a rigged survey. The advertiser of brand A, for example, might give out free samples of his product—and then follow up this move with a street-corner survey asking consumers which brand they prefer. Obviously, more than the normal number of people are going to mention brand A—if only because it is freshest in their minds.

The meaningless statement is another advertising ploy. "Our brand is better," reads the typical ad of this type. But better than what? How much better? Why better? The advertiser convenient-

"While your scientific integrity is commendable, Hopkins, I might remind you that around here Brand X never tests out ahead of our product." (© *The Wall Street Journal.*)

ly forgets to answer these pertinent questions—for all he wants to do is to leave the reader or listener with a positive feeling about his product.

The old "lowest cost" claim would fall into the same category. The problem, of course, is that the word "cost" is never defined. Does it mean the lowest initial outlay, the lowest operating cost, the lowest repair cost, or just what? Certainly the advertiser doesn't tell us—and with good reason. What he is trying to do is confuse cost with value. You can buy the cheapest appliance, but it may well break down the next day. Is this then your best choice? Obviously not—and in terms of value received it may well be the most expensive.

Relevancy is another common advertising deception. One cigarette may have less tar or nicotine than another. But is the difference really significant? Then, too, even if it is, does it make any significant difference as far as health hazards are concerned? Again none of these relevant questions ever seem to be answered by the advertiser.

Another illustration of the many irrelevancies that crop up almost daily in advertising: In early 1973 a large automobile company came out with a full-page advertisement in many of the nation's newspapers protesting the 1975–1976 federal emissions standards for automobiles. It stated, "The vegetation in your backyard gives off as many hydrocarbons as the 1975–76 law permits your car." The obvious implication: the furor over auto pollution was a tempest in a teapot.

While it may indeed be true that vegetation emissions might equal auto emissions, the ad stopped far short of the whole truth. The hydrocarbons produced by cars include compounds which are carcinogenic (cause cancer) in animals and which have been implicated as a cause of human cancer. On the other hand, the hydrocarbon emissions from vegetation are of a harmless sort—and in no way have they been found to be related to the hazardous types spewed out by cars. To imply they're the same is little more than outright deceit.

The examples could go on and on. But let's look at one more popular distortion before leaving the subject. This one involves the speed at which a proprietary drug will work on a patient. Perhaps it is true, as claimed, that tablet A does dissolve 2.5 seconds faster than tablet B. But does this mean that tablet A is more effective? In short, is it the speed at which a tablet dissolves, its chemical composition, or perhaps a third factor that does the required job? The 2.5-second dissolving differential certainly doesn't give the answer. But it does provide a specific number—suggesting that in some way tablet A has been "proven" better. Unfortunately, such spurious "proof" is often enough to swing the odds in favor of tablet A.

One of the problems in halting some of the above practices is the fact that truth seldom pays off where advertising is concerned. A study released in early 1973, for example, found that deceitful ads can be far more persuasive than promotions that tell the simple truth.

Specifically, a panel of 100 largely middle-income consumers watched the truthful commercials, and another group of the same size, income, and educational level saw the dishonest versions. Both sets of commercials used the same actors and, except for the misleading bits, the same language. Yet in four of the six tests, the cheating commercial placed well ahead of the honest promotion in coaxing the audience into a buying mood.

Two of the commercials quoted which brought a strong positive consumer response: (1) a mythical Pro Gro plant fertilizer which stressed that it contained protein—although protein is essentially irrelevant for plant growth; (2) D-Corn, a bunion remedy which was touted to contain four times as much methyl-glyoxal as its competitor; yet no evidence was offered to support the implication that increasing the amount of this chemical might speed corn removal.

Given the above results, it is easy to see why advertisers resist any and all attempts at what for want of a better word might be called a truth-in-advertising code.

KEEPING INDUSTRY HONEST

Can anything be done about all these spurious advertisements? Up until recently the answer would have had to be, No. But over the past few years some halting attempts to clean up the most blatant numerical abuses have been initiated. This in itself has to be regarded as encouraging; for the first steps toward any cure involve identification and isolation of the problem areas.

The Federal Trade Commission (FTC) has been spearheading the new drive. The agency made its first big move in 1971, when it asked members of selected consumer goods industries to (1) submit data substantiating their numerical performance claims and (2) make this substantiating material available to the public.

The program's goals were twofold: education and deterrence. It was hoped that public disclosure of the substantiating material, or its lack, would assist consumers both to evaluate claims and to make a rational choice between those of competing companies. At the same time, it was believed that public scrutiny of the substantiating data would encourage advertisers to eschew their former practices of making unsubstantiated claims.

The Commission's preliminary analysis in 1972 yielded a number of general conclusions. First and foremost, it was found that the quality of substantiation varied greatly from company to company and from claim to claim. Many claims were documented in a most impressive manner; on the other hand, serious questions as to the adequacy of data used to support the claims they purported to document arose in about 30 percent of the responses.

Also, with respect to those claims that were made, it was found that in at least another 30 percent of the cases the substantiating material was so technical in nature that it apparently required special expertise, far beyond the capacity of the average consumer, to evaluate.

Sharp industry differences also turned up, with the study making it quite clear that certain sectors left much to be desired when it came to making advertising claims. A case in point: the automobile industry. Serious doubts were expressed over claims involving fuel economy, economy of maintenance, passenger comfort, acceleration, braking, emission control and safety, handling, new features, engine size and horsepower, steering, economy of purchase, resale value, durability, and frequency of repair. Each of these claims was considered by the FTC to be susceptible to objective numerical verification.

The results: Although each manufacturer made some response

to every item, the FTC found that a good portion of the material submitted as the sole support for a specific claim bore little apparent relevance to the claim in question. And in many cases, even where empirical data in support of the claims were furnished, the Commission felt they were not sufficiently complete. Then, too, in a few cases, the manufacturers relied solely on their own unsupported assertions as substantiation for a claim.

In still other instances, the submitted documentation (although apparently substantiating the advertising claim and relatively complete) did not appear to reflect consumer experience. In short, the relevancy of the documentation was challenged. Gas mileage claims were singled out on this score.

Specifically, most of the tests appeared to have been performed by professional drivers adhering to rigorous test standards — standards which were hardly representative of normal consumer driving conditions.

Armed with this discovery, the FTC then asked various automotive magazines about the industry's testing methods. The agency was informed that all manufacturers rely on "optimum test conditions" — conditions that are unlikely to be duplicated by untrained consumers driving under typical driving conditions. Ergo, the tests and claims which they supported were deemed to be of questionable relevance to the prospective purchaser of the vehicle. Moreover, since the submissions as a whole reflected no standardized test procedure, it would have been quite difficult, if not impossible, to compare mileage claims made by different manufacturers — or to extrapolate from these claims to normal consumer driving experience.

In addition to the responses which, for various reasons, were considered questionable, there remained a goodly portion that were so technical in nature that they could not readily be understood or evaluated by persons lacking technical training and experience. This problem is noted not to criticize the substantiation, but to make clear that any ad-substantiation program directed toward placing data on a public record must be judged in light of the fact that consumers will often not be able to comprehend the data. Similarly, descriptions of some automotive parts sometimes employed technical language that was incomprehensible to the Commission staff.

Advertising substantiation, of course, isn't the only FTC approach being pursued. Another promising one: "Corrective" advertising. In 1972, for example, the agency forced two sugar trade associations—Sugar Association, Inc., and Sugar Information, Inc.—to agree to take full-page ads in seven national magazines saying that they didn't have any substantiation for their claims that sugar helps people lose weight.

The trade associations, which represent sugar processors, growers, and refiners, had to run ads stating that their earlier suggestion that taking sugar before meals would help curb appetite isn't a "magic formula for losing weight."

The advertisements ran under the headline, "The plain truth about your sweet tooth." They began, "Do you recall the messages we brought you in the past about sugar? How something with sugar in it before meals could help you curb appetite? We hope you didn't get the idea that our little diet tip was any magic formula for losing weight."

The ad went on to state that "research hasn't established that consuming sugar before meals will contribute to weight reduction or even keep you from gaining weight."

This kind of "eating crow" could have a substantial impact on advertisers. Moreover, because of its unique message, it's bound to capture the public's eye, thereby enhancing its effectiveness.

CAUSE AND EFFECT

Here is where the old "smoking and cancer" controversy comes to the fore. When two variables are statistically related, there is a general tendency to link the two together—to imply that one causes the other. The only problem is that statistical relationship in and of itself does not prove causality. The price of eggs in New York might vary with the rainfall in New Guinea. But it would be the height of folly to suggest a New Guinea rainfall–New York egg price cause-and-effect relationship. In short, causality needs more than just numbers. It needs strong confirmation either in the laboratory or by established theory.

This can perhaps best be understood by zeroing in on the cigarette-cancer controversy. No one, of course, would deny that there is a significant statistical relationship between these two variables. But that's it. True, many jump the gun and say that if the difference between smokers and nonsmokers is statistically significant, then smoking must be a cancer-causing agent.

Spokesmen for the tobacco industry, however, have come up with a cogent rebuttal. They admit that more smokers than nonsmokers have been hit with the dreaded disease. But they say this proves nothing. Perhaps, they argue, a third factor, such as tension, might make more people prone to both smoking and cancer.

And the argument is persuasive up to a point, for if tension leads to smoking, who is to say whether or not it can also cause cancer? Looking at this another way, the fact that cancer and smoking move in tandem proves nothing if both are, indeed, "caused" by a third variable.

But the fact that the existence of a statistical relationship doesn't prove cause and effect doesn't mean that we should ignore it, either. In many cases it has been the starting point for further research. This is exactly what happened in the smoking area. Once the statistical relationship was established, extensive laboratory experiments were initiated. And these have shown beyond a shadow of a doubt that certain chemicals in tobacco can physically cause cancer. No wonder then that the government now requires a warning on all cigarette packages stating that "smoking may be injurious to your health." No wonder,

too, that there is now a concerted effort on the part of many to
"kick the habit."

As one wag, impressed with the numerical relationship between smoking and cancer put it, "Where there's smoke, there's
fire."

Statistical relationships pointing the way toward theory or
laboratory verification of cause and effect need not be limited
to the disease sphere. In economics much of the current theory
of the business cycle has been based on observations relating
spending to income, and investment to rate of return. Then,
too, many of our agricultural advances owe their initial start to
observed statistical relationships. Such relationships were then
taken into the laboratory for testing. Result: an agricultural
revolution that has doubled the world's food supply in a few
decades.

The lesson in all cases is the same: Statistical relationships
do not establish cause and effect, but they do suggest the possibility of such causality. In a sense, they provide some of the
indispensable clues to a better understanding of our social and
physical worlds.

SEVEN

THE RISKY SAMPLE

The cook sipping the soup. The commuter looking out over the highway before choosing his route to work. The market researcher asking you whether you prefer brand A or brand B.

Seemingly unrelated incidents? Sure. But there's a common thread running through all these actions: sampling. All three people are trying to make an overall appraisal on the basis of a relatively limited amount of information.

The cook, by tasting one spoonful of the soup, can pretty well infer what the whole pot of soup will taste like. The commuter, when he sees traffic crawling along the highway, concludes that it will be that way all the way into town, and so he elects to take the old county route instead. And the market researcher, who finds that 70 out of 100 people queried prefer brand A, might conclude that 70 percent of the entire population prefer brand A.

In a sense, then, we all put our trust in the sampling method. Indeed, the technique of making an estimate of the characteristics of a large mass of data by examining a sample is an old and intuitive one. But there is one big difference between the sampling done by the cook and that done by the commuter and market researcher.

The cook is pretty sure of her inference. Not so the latter two samplers. The traffic snarl may clear up around the next bend in the road—and then again it may not. Similarly, the next 100 people queried by the market researcher may, or may not, have the same opinions as the first 100 people queried.

Why is the cook so sure, while the other two have their doubts? It's simply because the cook is sampling a homogeneous product—that is, a product subject to very little variability. On the other hand, the other two samples are not necessarily homogeneous. Traffic on the section of the highway being viewed is not necessarily the same as on every other section. And the second 100 people interviewed may not be of the same opinion as the first 100.

Therein lies the problem of sampling social, business, and economic data. We can never be sure when we view just one small segment of the population—or, as statisticians like to call it, of the "universe."

As such it's impossible to say with certainty that 70 percent of the people prefer brand A—although this is what is always inferred by people quoting such statistics. Perhaps the sampler just happened to pick the only 70 people in the area who preferred brand A. Highly unlikely—but still within the realm of possibility or chance.

In short, in the typical sampling problem, one can never be sure.

That's not to imply, of course, that sampling is useless. Quite the contrary, for this approach is extremely useful, both in terms of cutting costs and in terms of coming up with an educated guess where no other estimating procedure is possible. Thus we all want to know how consumer prices fare every month. But if Uncle Sam were to monitor each and every supermarket transaction in the nation, the answer wouldn't be available for years—at which time it would be virtually valueless for consumers and government policy makers alike.

The thing to remember, however, is that every time we sample, there are risks involved. There's no guarantee that the sample will be 100 percent correct. The problem for the recipient of the sample figure, then, is to assess the chance of error, and then make a decision—realizing that there's some chance that it could be the wrong one.

This can best be illustrated by the manufacturer who samples, say, 1 percent of the widgets coming off a production line. This sample might let us know with about 99 percent accuracy whether to accept or reject the lot. The businessman might well be willing to accept the 1 percent risk of making the wrong decision in return for the tremendous potential savings derived from being able to differentiate most of the time between good and bad lots. It's a risk, yes. But it's a calculated risk—one that is likely to pay off over the longer pull, with the number of times the analyst guesses right for outweighing the one or two times he makes the wrong decision.

The remainder of this chapter will be devoted to the evaluation of such risk, and how, by careful sample selection, this risk can be minimized.

CHOOSING A REPRESENTATIVE SAMPLE

One of the most common causes of sampling error stems from poor survey techniques. No matter how big the sample or how scrupulously it is processed, it is virtually worthless unless it truly represents the population from which it is drawn.

"Representativeness" is, thus, a sampling must. Unless the sample is a mirror image of the true population or universe, bias and error are likely to creep into your results—and may well lead to the wrong conclusions.

If nothing else, be wary of "street corner" research. That's been the big mistake of so many urban sociologists. One of their big errors: trying to study the slum by means of direct observation. It is true that the researcher using this approach sees more action at close range. Yet street corner studies have bias built into them. They fail precisely at getting the representativeness that every sample needs if it's to be a valid one.

Street corner observers can get to know only a limited number of people. Moreover, only those people who will make themselves available for the heavy exchange of conversation vital to the researcher's talks will be studied. But the sad fact is that today the most accessible members of low-income neighborhoods tend to be the least well controlled and the least respectable members—hardly a true cross section of the slum being studied.

In short, this kind of sample does not paint a truly representative picture of the sentiments and characteristics of the slum community. Ergo, you read almost daily about the anger and frustration of the ghetto but very little about the fear—the fear that stems directly from the anger, violence, and crime of the few upon whom the interviewer focuses.

Sometimes the failure to achieve representativeness is a bit more subtle. Many times, for example, the survey taker forgets one obvious fact (a fact that is obvious to everybody but him) and follows through all the way down to final publication— only to find that this one "mistake" has thrown his whole conclusion out of kilter.

The most horrendous—and probably the most infamous— example of this occurred back in the 1930s, when sampling was just coming into its own. The time was late 1935. The opinion ostensibly being measured: the voter's choice for the president—the incumbent, President Roosevelt, or the challenger, Senator Alfred M. Landon of Kansas. Everything was planned impeccably—with a statistically significant number of voters to be called up from all sections of the country. But the planners forgot one basic fact: the use of the phone itself was introducing a bias into the sample. Remember, this was 1935, and the people who owned phones at that time did not represent a cross section of the American public. Quite the contrary. Telephones were a luxury then—and the people being sampled were the relatively affluent ones—and hence the ones more likely to vote for the Republican candidate, Alfred M. Landon.

The results were predictable. The poll (taken by the *Literary Digest Magazine*) suggested that Landon would be a shoo-in. The actual results are history: President Roosevelt won a second term by a landslide.

Hindsight, of course, nearly always reveals where the mistake has occurred. The problem is to be aware of all the potential pitfalls so that such mistakes do not take place. Sometimes these are pretty hard to spot.

Take the seemingly simple question of measuring the number of customers who might patronize a store at a potential site. If your survey is taken during the morning, you may get one conclusion. If it is taken at noon or during the evening rush hour, completely different sets of conclusions may come out of the survey.

Summing up, a sample must be a miniature snapshot of the real world. Any distortion will be magnified—and hence could be an open invitation to disaster. To achieve such representativeness, a sample should be a random one—that is, each item in the population being measured should have an equal chance of being chosen or drawn.

But sometimes complications arise even with this seemingly

simple concept. Thus to take care of any heterogeneous aspect, it is often necessary to "stratify" the random sample—ergo, the name "stratified" sample. The approach here is to divide the population into strata or levels, each of which is more or less homogeneous, and then take random samples from each stratum. If, for example, the problem is to make a study of department store sales, the sample might be divided into strata consisting of stores grossing over $10 million, stores grossing $5 million to $10 million, etc. A random sample would be drawn from each stratum and then all would be combined with proper weighting into an estimate for the population. It follows, then, that with more accurate estimates for each stratum, the estimate of the entire population should be more accurate for a given sample size.

There are several variations of this basic approach—all aimed at making the sampling easier or more reliable and accurate. One of these, called *sequential sampling,* actually involves the taking of a relatively small sample—and then enlarging on it until a statistically reliable result can be achieved. It can be shown that this approach, involving a given number of sampled items, yields more acceptable results than one large sample of the same size.

Sometimes what may appear on the surface as being representative (even to the extent that the sampling was random) may in actual fact be not representative. The problem and attending distortions often arise when mail surveys are attempted.

Specifically, a representative cross section of the population may be queried, and yet the results do not represent the population. That's because the people who might normally respond constitute a different cross section from those who might choose to ignore the survey. Thus the proportion of responses by sex, ethnic group, age, income, etc., may not be in line with the proportions of those factors in the overall population.

Again it all boils down to the failure to obtain a representative sample—something which, more times than not, will bias your results. The problem isn't going unnoticed. Thus people who conduct mail surveys are now attempting to (1) increase the response of reluctant subgroups and (2) make corrections in their tallies when nonrepresentativeness persists.

GETTING PERTINENT ANSWERS

How to word your questionnaire is another formidable problem faced by surveyors. Sometimes a slight change can bring about a sharply different response. The designing of questions can also affect the sampler's interpretation. If the questions yield fuzzy answers, then the interpretation is likely to be fuzzy. Finally, care must be taken to see that the people who make the decision rather than those who use the product or service get the sample questionnaire. Examining the latter problem first:

1. *The wrong respondents.* It may seem hard to believe, but in many instances the wrong people are being asked the questions. The following should serve to illustrate this pitfall.

A state highway authority, wanting to determine the likely traffic density generated by a proposed toll route, decided to get the answer by conducting a survey of motorists (potential users of the road). The sampling technique was relatively straightforward: drivers of private automobiles as well as trucks in the area were asked whether they would use the new turnpike, given a specified schedule of tolls. Responses from both private motorists and truck drivers were so strongly in the affirmative that the proposed tolls were put into effect when the turnpike was opened.

But, lo and behold, the forecasted revenues failed to materialize. The big shortfall was in the truck area. Specifically, trucks (survey findings to the contrary) were just not using the toll road in the volumes predicted by the survey.

A closer look revealed where the samplers went off the track. The truck drivers who said they would use the toll route were obviously voicing their true preference (the route would be a lot easier and faster). But what the truck drivers and surveyors failed to reckon with was the fact that the drivers did not control their routings—instead they had to follow routes laid out for them by dispatchers at their home terminals.

Obviously, then, the survey should have hit dispatchers rather than truck drivers. In other words, the wrong people were being queried.

But the story has a happy ending. The turnpike people went to the dispatchers and emphasized the fact that the time-distance savings of the turnpike more than offset the cost of tolls. Their argument was persuasive. The turnpike quickly developed a healthy volume of truck traffic and moved from the red into the black.

2. *Poorly worded questions.* Here, too, serious problems can arise. In general, surveys should be constructed so that the respondent has little doubt as to what the questioner really means. To assure consistent results when asking for age, for example, it is usually a good idea to qualify the query with, say, a request that the respondent round out his answer to the nearest year. In short, don't be ambiguous.

At times, this ambiguity can lead to rather funny responses. For example, one standard navy form for recruits into the women's auxiliary contained a space marked "sex." One truthful, but confused, young lady answered: "Once, in Texas."

An even funnier response came to the Agricultural Department from a farmer who was asking for compensation for a cow killed by a Department truck. The farmer, coming upon the question, "Disposition of the dead cow" replied: "Kind and Generous."

It is also a good idea to avoid asking for "yes" or "no" responses. Sometimes they are loaded questions, as in the case of,

"Have you stopped beating your wife yet?" But more important, they often make it difficult for the respondent to give a meaningful answer. Suppose you are asked whether you smoke. Perhaps you do, but only occasionally. How do you answer? A better approach might be to ask the question this way: "Do you smoke (a) heavily, (b) moderately, (c) occasionally, or (d) never?"

3. *Questionable responses.* Many times the response received is untruthful—either because the respondent willfully lies or because the answer has been "suggested" to him. Ask a woman her age, or anyone whether or not he uses "soft" drugs, and a deceptive answer is more than likely.

At other times the question itself invites a distorted answer. Thus a question asking the housewife whether she would like a given product prepackaged is likely to evoke a "yes" answer, a response that may not materialize when she finds out how much more she would have to pay for this convenience. A better question might be: "Would you be willing to pay 10 cents extra for the prepackaged version of the product?"

4. *Limited analytical value.* Questions should be so worded as to wring out the last ounce of useful information. This then presupposes queries that permit classification, summation, and analysis. On the first score, responses must be classified and tabulated in a meaningful manner. This would usually eliminate the "Why do you smoke brand A?" type of question. A much better way to get at the underlying reasons is to submit a list of possible factors and then have the respondent check off the appropriate ones.

Once this is accomplished, the next step is to take all the properly classified responses and convert them into some action-oriented yardstick. Thus the Michigan Survey Research Center prepares an index of consumer confidence every three months on the basis of a series of questions. Without its index the reader would be hard put to evaluate the overall responses in one period as contrasted with those in another period.

By now it should be clear that the purely physical problem of working up a representative sample presents formidable obstacles. It can be done—and is being done. But it takes time, money, and a lot of savvy. Ignore market makeup, survey design, the type of questions asked—or any one of the host of other considerations discussed above—and your sample results are likely to show serious bias.

Such bias is of more than academic interest, too. In cases of business, it can result in the loss of considerable sums of money. In the case of public attitudes, such biased survey findings can suggest policies which in the long run may prove questionable and in some cases even disastrous.

ERRORS OF CHANCE

So far only avoidable mistakes have been discussed—those that could conceivably be eliminated by correct sampling design. But beyond this, there is yet another type of error present—the one due to the inherent nature of the sampling process itself. Specifically, on the basis of pure chance, we may pick a sample that differs significantly from the population or "universe."

At this point it might be well to illustrate the difference between these two types of sampling error with a simple example. The problem: You are a manufacturer of light bulbs and want to know how long your typical or average bulb will last before burning out.

The first type of error—the type discussed in the previous section—might occur if you took all your sample bulbs off one machine—or perhaps took them during late afternoon when worker fatigue might affect quality.

The second type of error—the chance or probability error, which is being discussed in this section—would occur no matter how carefully you selected the sample. It would occur because there is inherent variability in all production, and it would indeed be surprising if the first bulb sampled from a purely random sample would last as long as the second or third bulb from this same sample.

Following through on this, it is then quite conceivable that all of the bulbs in your random sample may turn out to have an average life somewhat different from the average life of your population (your entire production) of light bulbs.

There is little you can do about this second type of chance variation—except, of course, to recognize it, evaluate it, and qualify your conclusions with the statement that your results may vary from the real population by such and such a percentage.

Thus on the basis of statistical probability you might say that with a random sample of x bulbs with a normal life variation of y hours per bulb, you would expect your sample average to vary from the true population average by, say, about 5 percent.

Further qualifications would also be in order. You could make the above statement only with given confidence or probability. Thus your really correct statement would have to read: "A sample of x items with given y variation will vary about 5 percent from the true population 99 times out of 100."

Sounds complicated. But the fact is statistical probability theory is complicated. Nevertheless, it can be mastered. More important, it has to be mastered if meaningful sample evaluation is to be made. Equally significant, once sampling error is recognized, there are many different types of problems to which it can be applied. Again harking back to our bulb problem: A sample from manufacturer A may suggest a bulb life of 1,000 hours while a sample from manufacturer B may suggest a life of 1,050 hours. The question: Is the difference significant enough to warrant our shifting over from manufacturer A to manufacturer B?

Again probability theory comes to the rescue. Given sample size and variability, we can come up with the odds that the variation between the two manufacturers is due to chance.

If the chance odds are large (say, 80 percent), it may not pay to shift. On the other hand, if the odds are small (say, 20 percent), then we may say there is a significant difference between the bulbs of the two manufacturers and it indeed may be worthwhile to consider a shift in our purchasing strategy.

Probability also plays a role in correlation. Take the discussion of the previous chapter on the relationship between cigarette smoking and cancer. We may note a close relationship, but only probability can tell us what the odds are that this relationship is significant or that it is due to chance. In this case, of course, the correlation was found to be quite significant. In some studies the possibility of a chance relationship was found to be under 1 percent. And this then gave the researchers the justification for embarking on a crash laboratory program to see if they could establish true causality.

This kind of "odds fixing" is also quite important in brand preference studies. Thus if 40 percent prefer brand A and 60 percent prefer brand B, we can with the appropriate information determine whether such a difference is statistically significant. But again the answer is always in terms of odds—never in terms of 100 percent certainty.

There's still another plus to sampling that deserves honorable mention: It is often the only way of getting an answer. If we were to test all light bulbs to ascertain their longevity, we would have none left to sell. In other words, where destructive testing is involved, sampling is the only way out.

Before leaving the subject, it might be well to point out that while all the above applications require statistical savvy, the basic probability theory upon which it is all based is really quite simple.

The classical coin-tossing problem can best illustrate the principle. Everybody knows that the odds of getting a head or a tail are approximately 50 to 50 (unless, of course, the coin is loaded).

Carrying this one step further, we would expect that if we

tossed two coins, we would end up with one head and one tail.

Now actually toss two coins several times. A fair amount of times, you will get your one head and one tail. But in more than a few cases you will wind up with either two heads or two tails. So we never know with certainty that we will end up with one head and one tail.

Statisticians can, then, calculate the odds of not getting the one head—one tail result. They find that there's a 25 percent chance of getting either two heads or two tails. Here's how: They say the probability of getting one head when tossing the first coin is one-half. Then they say the odds of getting a head on the second coin is also one-half. The odds of getting two heads are then calculated by multiplying the odds together: $\frac{1}{2} \times \frac{1}{2} = \frac{1}{4}$.

For those who would argue that the odds should be added together, consider that such a procedure would yield a 100 percent chance of getting two heads ($\frac{1}{2} + \frac{1}{2} = 1$). This would automatically preclude the possibility of obtaining two tails—or the even more likely probability of the one-head-and-one-tail combination. Obviously, then, multiplication rather than addition of the odds is what is required.

For those interested, the true probabilities for this coin problem come out: $\frac{1}{4}$ or 25 percent for two heads; $\frac{1}{2}$ or 50 percent for one head and one tail; and $\frac{1}{4}$ or 25 percent for two tails.

This seems like a laborious process—and it is. But there's a neat formula—the so-called binomial theorem—which gives the answers in a matter of seconds for any number of coins being tossed.

The point to remember: Whether you sample coins, light bulbs, or consumer opinions, there is always a possibility of variations. As such, when sampling is involved, results must always be couched in terms of probability or odds—primarily because chance factors are always operating when only a small portion of the population or universe is being tested.

WHAT SIZE SAMPLE?

The odds of making an error can, of course, be cut down by increasing the size of the sample. This is little more than common sense. For, as a sample size approaches the population, we are relying less and less on sampling and more and more on actual enumeration.

Indeed, statistically, it can be shown that accuracy tends to increase with the square root of the sample size. If, for example, you want to double accuracy, the sample would have to be increased fourfold. The point here is that we soon begin to reach a point of diminishing returns. Thus a 25-fold increase in sample size would be needed to yield a fivefold increase in accuracy.

So there are limits. Also, as pointed out earlier, 100 percent accuracy is often not needed. Clearly, in terms of cost or practicality, it may even be undesirable. Thus the problem is to determine the trade-off between cost and practicality on the one hand and accuracy on the other.

Once the degree of accuracy desired or required is determined, it is necessary to pick the sample size that will yield this chosen degree of precision. In a sense, the decision on the exact amount of precision needed is a subjective one. A parachute maker would demand almost 100 percent precision. On the other hand, our bulb maker can settle for something less. Money is still another factor. Thus, where millions of dollars are riding on a decision, we would want a lot more precision than where only a few dollars are involved. There are no statistical rules here. You weigh the risks against sampling cost and make your decision accordingly.

Risk isn't the only factor that should determine sample size. The variability of the product being tested is another key factor. Thus sampling from a homogeneous rich or homogeneous poor community would require a smaller sample than sampling from a mixed community of the same size because variation in the former would be smaller than in the latter. In a way, this is little more than common sense. If everybody is alike, then why look at everybody to get a profile of the group?

Strange as it may seem, the value of the population also plays a role on how big a sample may be needed. If you're sampling to find the number of items that are defective in a large lot, then a lot with a population of 10 percent defective would require a larger sample than, say, a lot with only 2 percent defective.

Equally strange, the population size has little or no influence on accuracy. Thus whether the population is 10,000 or 100,000, you can probably choose the same sample size and come up with the same degree of accuracy.

It is difficult to give common-sense reasons why the value of the population does affect sample size while the size of the population doesn't. Suffice to say, however, statistical formulas give ample justification for these conclusions.

Before leaving the subject, it might also be pointed out that sample size can also affect the reliability and accuracy of relationships between two magnitudes or, as the statisticians like to call them, variables. It can be shown statistically, for example, that spending is correlated with income—the higher the income, the greater the spending. But how reliable is the relationship? Again chance factors are operating—and again, the larger the number of observations, the more confident we can be in the observed correlation.

That's why, for example, the medical profession, in studying the relationship between cholesterol intake and heart attacks— or smoking and cancer—insists upon a relatively large sample size. For too small a sample would leave their findings subject to relatively wide chance variation—and hence considerable doubt.

Trend projections, too, are open to chance variation. A given 5 percent growth rate based on, say, 6 years would be less reliable than the same 5 percent growth rate based on 10 or 12 years. Again, the larger the number of observations, the smaller the chance influence and the greater the confidence we can have in our projection.

Upshot: Beware of correlation or trends based on small samples. They may be valid—but then again, they may not be. To make sure, ask about the "significance" in the case of correlation or the "standard error" (probable deviation) in the case of a projected trend.

EIGHT

WATCH THOSE ASSUMPTIONS

We've all had the rug pulled out from under us at one time or another.

The expectation of a raise nixed because company business took a tumble, the four-week vacation postponed because your assistant suddenly quit, or even the Sunday picnic scratched because of rain—all these have one thing in common: a shift in underlying assumptions which makes the basic premise (a raise, vacation, picnic, etc.) untenable.

The problem of these assumptions becomes especially acute when numbers or projections are involved. For every such number is usually premised on certain underlying ground rules. Change the ground rules, and the forecast may have to be changed.

The problem in such cases, of course, is that the number of

assumptions must necessarily be quite large. Estimates of food prices next year, for example, will depend upon weather, government price supports, farmers' planting decisions, market demand, exports, etc. Change any of these, and prices automatically change.

Thus in late 1972 bread prices suddenly and unexpectedly rose, with the impetus coming from a surprise export wheat deal with the Russians. The sale called for shipment of large tonnages of wheat within a matter of just a few months. The inevitable effect: rising wheat quotes and, hence, bread prices.

Now nobody could have predicted that this would occur—and even if they had, the timing and magnitude of the transaction could in no way have been foreseen in advance. The lesson is clear: There is no way of guaranteeing that yesterday's assumptions will be valid tomorrow or even today.

Even reasonable assumptions are not always tenable. Forecasts involving sales of seasonal goods are usually premised on the likelihood of normal weather conditions. Let the temperature or precipitation vary significantly from the norm, and the sales of the goods in question can change substantially. More about this below.

Time also has a habit of making a seemingly infallible assumption obsolete. Tastes and habits change—and what might have been virtually 100 percent certain a few years ago may today be on pretty shaky ground. Examples of this discussed below include (1) the percent of income spent on consumer durables and (2) the product mix of the market basket of goods. On the latter score there has been a marked shift—away from expensive, frilly models and toward the cheaper, more functional types.

In short, many a seemingly perfect forecast has come crashing down because of the shifting sands on which its foundation was laid. Assume prosperity for the next year, and no matter how bullish your own firm's outlook is, sales will be lower than anticipated if the economy should go into a tailspin.

UNTENABLE ASSUMPTIONS

More times than they like to admit, consumers are guilty of making what eventually turn out to be unwarranted assumptions. Take the case of the auto buyer who sees his favorite car model advertised in the newspaper for $2,500. Checking his bankbook, he figures he can just about make it. And naïvely relying on the advertised $2,500 figure, he rushes down to the nearest showroom.

But when he gets there, he's in for the surprise of his life. First, he finds that for the $2,500 price he may not be entitled to a radio, heater, air conditioner, power steering, or any one of the other "extras" that he has become accustomed to. More likely than not, this "optional" equipment will usually add $700 or more to the price of the average vehicle.

Then, too, the consumer may have figured that the price included delivery charges and sales taxes. No such luck, for the original quote in the newspaper covered only the bare-bones car—freight and sales taxes are extras. For vehicles in many states, these nuisance extras can add up to another $200 or so. In short, John Q. Public, to actually take possession of the car he really wants, would have to spend a lot more than the sum suggested by the eye-catching $2,500 price tag. Specifically, he'd probably have to pay close to a $1,000 or 40 percent more.

The lesson is clear: Don't blithely assume that the price you see advertised—for a car or any other big-ticket consumer item—will be the amount of money you'll actually be shelling out. Advertisers have a vested interest in "hooking" you with a low-sounding price. Indeed, that's the name of the game as far as Madison Avenue hucksters are concerned. And in a sense they really haven't lied. All they've done is to conveniently forget to tell you about other expenses which will most surely add to your overall purchase outlay.

Nor are businessmen exempt either from the untenable-assumption syndrome. Take the example of company X, which a few years ago tried to break into the lucrative garden equipment market. Their new line, while hardly cheaper than existing lines, was expected to be a major success because their corporate name was well-known and because they felt their equipment was better-made. As such, they expected to capture about 10 percent of the market in about two years.

But much to their chagrin, the project turned out to be a major disappointment. Again their assumptions were at fault. First, the "gut" feeling that company X's name in other fields would carry over into the garden equipment area was completely unwarranted. The other firms over the past decade had built up a loyal following, and their customers weren't about to desert to a newcomer.

Their second assumption, that their "better-made" product would carry the day for them, also turned out to be unwarranted. Home owners had no way of verifying the apparently truthful claim of superiority. Moreover, they weren't particularly impressed, either—having heard the same pitch from hundreds of other manufacturers in many other different areas.

Faced with such insurmountable problems, company X's new program proved a bust—despite the fact that the garden implement market grew strongly over the period (just about in line with what the company X's market research people had projected).

Finally, even the government is often guilty of making unwarranted assumptions. A good example of this occurred a few years ago when Uncle Sam's economists thought business was growing too fast—and, therefore, wanted to press down on the spending brakes.

Their goal: A slowdown in consumer outlays. Their technique: the imposition of a 10 percent income tax surcharge. The theory:

Such a surcharge would reduce spendable income, thereby inducing families to cut back on their purchases of luxuries such as cars and appliances.

Unfortunately none of this seemingly impeccable strategy panned out. In a strange reversal of past experience, people actually began to accelerate their spending on such goods. Hindsight reveals why. The assumption of a drop-off in spending failed to take into account special psychological factors which were operating at that time. Specifically, there had been talk of a surcharge for nearly half a year before its actual imposition. This had so unnerved consumers that they had already cut back on spending and when the blow actually fell there was a feeling of relief.

This was reinforced by the fact that the consumer interpretation of the surtax had been faulty. Many had thought their tax rate was going up by 10 percent. But it was only their tax bill which went up by this amount.

The difference is significant. Thus a wage earner who paid $2,000 of his $10,000 income to Uncle Sam felt his rate would be going up from 20 percent ($2,000) to 30 percent ($3,000). In actual fact, however, he was now being asked to pay 10 percent more on his $2,000 tax bill, or only $2,200 in total. In other words, his rate went up only 2 percentage points—from 20 percent to 22 percent.

Result: People were pleasantly surprised over the small amount of additional income that was being withheld. They felt so relieved about this that they eased up and began to spend more freely.

Another factor may have been operating at that time too. The surcharge weighed most heavily on the upper income levels— those taxpayers who had enough discretionary income to offset the tax bite by saving less.

On the other hand, the poor and lower-middle classes were hit only slightly or not at all. Since these people spend all they take in, their spending also remained on a high level.

The need to spare the poor was of course a social necessity — and few would suggest anything else. But from a purely economic point of view, a tax that weighed more heavily on the poor would have been more effective in curtailing consumer spending.

Sometimes unrealistic hypotheses are embraced because we want to fool ourselves. Assuming the smaller-than-realistic new-car outlay as described above, even though we really know better, the idea of a low price will give us the excuse we need to go down to the auto showroom. Similarly, in planning a vacation in Europe next year, more than one family has conveniently forgotten to factor in contingency expenditures (illness, etc.) — despite the fact that such expenditures invariably crop up over the course of a year. The end result: costly cancellations or living beyond one's means, neither of which is particularly desirable.

CHANGING ASSUMPTIONS

In a great many other cases, assumptions may be perfectly valid at the time they are made. The only trouble is that things very rarely stand still. Many times it's simply an act of God or nature that turns a particular assumption upside down.

This is particularly true where seasonal influences are operative. Whether it be sales of soft drinks, air conditioners, or the latest Easter outfit, company performance to a large extent is dependent upon climatic conditions. The problem is, however, that no one can predict weather with anything even resembling a fair amount of accuracy. So assumptions must necessarily involve weather patterns over recent years. And there's little to guarantee that such patterns will be repeated with regularity from year to year.

This problem is something that plagues air-conditioner makers more than anyone else, because of the large dollar value of the product and the long-lead times necessary for production. A cool summer can drop projected annual sales by as much as 20 percent and 30 percent, leaving producers with top-heavy inventories. On the other hand, a simmering heat wave can clear dealer shelves in a matter of days. The latter occurred during one recent year—with the industry unable to meet skyrocketing demand. The aggravating point about all this is that these sales could never be made up. When additional air conditioners came off the production line a month later, the heat wave, and hence demand, had subsided. The industry figured it had lost 10 percent of its potential because of its inability to meet the demand at the proper time. Their only hope was that consumers would remember— and that during the ensuing year buyers would buy early.

Of course, this hope never materialized. People have notoriously short memories, particularly where the layout of large sums of money is involved. And as might have been predicted, they waited until the next heat wave before they again showed any great interest in purchases of these $200-plus items.

Product-mix assumptions also have a habit of going awry. Will people buy top-of-the-line or bottom-of-the-line merchandise? Detroit has a particular problem on this score, with the auto industry hard put to estimate what portion of their car sales will be subcompact, compact, regular-size, or luxury vehicles. Indeed, in one recent year, the industry failed to take fully into account increasing price-consciousness and "trading down" to smaller and cheaper cars. Result: Companies could not meet demand and lost potential sales to foreign competitors who specialize in these lower priced cars. (More about the fickle consumer in the following section.)

Assuming that economic barometers measure the same phenomena today that they measured yesterday is also questionable in many instances. Unemployment statistics can best illustrate the pitfall involved here. It was pointed out in 1970 (when welfare reform legislation was being discussed) that jobless figures under a reformed system might mean something a lot different from what they did heretofore.

Reason: To be counted as unemployed, one not only has to be out of work but also has to be actively seeking work. Under the then-existing "welfare" approach, it was advantageous for the poor to say they were not looking for work—for if they were to get jobs, their welfare payments would be cut off. But consider the reform program, where benefits would have depended on a willingness to work. This would have persuaded many of the unemployed to say they were looking for work (even if they weren't looking very hard).

A switch to the latter response would automatically push these people into the ranks of the "statistically" unemployed. Socially it would, of course, be a step forward to bring any hidden unemployment into the open. But the higher unemployment figure, which could be political dynamite, would not mean that any bigger portion of our labor potential was being underutilized.

Long-range statistical projections, too, are particularly prone to errors stemming from changing assumptions or ground rules. Housing provides a perfect example of this. Thus in the late 1960s and early 1970s predictions based on need and capability called for the construction of more than 2 million housing units per year. This assumed that enough money to finance this number of starts would be available.

Unfortunately, it was at just about this time that the government stepped up its fight against inflation by making money both tighter and more expensive. This hurt housing and, as a result, housing starts during these years averaged below 1.5 million units, and in one year, during the height of the credit squeeze, they weren't much above 1 million units.

This effect, in turn, played hob with long-range forecasts of building material producers. Demand for cement, brick, glass, lumber, and a host of other items fell sharply below forecasted levels. Some of the firms involved were alert to the money shift and adjusted their forecasts accordingly. But others, forgetting to make periodic checkups of underlying assumptions, found themselves overproducing, with the resultant heavier-than-normal inventory buildup forcing a sharp drop in profits and margins.

Why were so many unable to shift gears during this period? Several reasons have been advanced.

First, there's the problem of laziness and/or fear. If conditions change, that could mean major recalculations—and maybe calling the boss's attention to the fact that you have changed your thinking on some crucial aspect of company sales. This latter factor could be unnerving—because, just as a forecaster doesn't like to have the rug pulled out from under him, neither does the production or sales manager.

These people, too, have made firm plans and are unhappy about seeing them changed. They may well react by castigating the forecaster for not being able to "make up his mind." But there's a good retort to this: Ground rules have changed—and he, the forecaster, is being paid to stay on top of just such shifts.

Some other forecasters try to get out of the box by attempting to convince themselves that any changes that have taken place are of minor importance. A few years back a sales tax was instituted in a large Eastern state. A big retail chain chose to ignore the effect of this tax on prices—and found out that many of its customers were now going across the state border, to the extent that the chain's annual sales fell several hundred million dollars below projected levels.

Finally there's the "closed-mind" syndrome. There are far too many people who, once they have made their projection, are uninterested in anything that subsequently happens. These people have done their job. If a strike or a flood occurs, let management make the necessary corrections. It's no longer their worry. This is "sloppy" workmanship in its worst form.

Finally, some assumption changes are ignored simply because the assumptions involved are implicit ones—and as such often go unrecognized when changes occur. Thus in 1972 a devaluation of the dollar was widely expected to improve the United States balance of trade—in part by discouraging imports, which would become more expensive because of the devaluation. But the actual results were quite different. Imports jumped by a whopping 22 percent that year.

What went wrong? The explicit assumption that import prices would rise did indeed turn out to be correct. But an implicit assumption that other determinants of imports would remain unchanged proved to be untenable. Specifically, 1972 turned out to be a boom year, and the extra demand generated by the boom more than offset any import loss engendered by higher prices. Had analysts recognized this implicit assumption, they would not have been quite as surprised when imports continued to skyrocket.

By this time it should be clear that changing assumptions constitute one of the major problems in maintaining statistical accuracy. The big problem is how to deal with them. Business men, in general, have found that major monthly or quarterly checkups are the answer. Then, too, more and more outfits now require that each report carry a detailed listing of the underlying assumptions that went into the numbers given. Not only are the explicit assumptions required to be spelled out, but also all of the implicit ones. In still other cases, different sets of projections may be presented—one under one set of assumptions, the other under another possible set.

Detroit, burnt so many times, has often tended to opt for this latter "dual forecast" approach. Whenever a tax change is being discussed, this industry is more than likely to come up with two different, equally likely, forecasts. One steel company did the same when import quotas and dollar devaluation were being considered. Reason: Either of these two moves would have had a major impact on their domestic sales.

Admittedly, the solution isn't quite that easy on the consumer level. Nevertheless it behooves every user of statistics to think about something more than just pure numbers or pure technique. The thing to remember: Every number is based on assumptions; therefore the validity of the number is only as good as the validity of the assumptions behind it.

THE FICKLE CONSUMER

John Q. Public seems to take almost fiendish delight in confounding the experts. Any assumption about what he may do has to be regarded as tentative at best. One example: the strange behavior of consumers during the income tax rise discussed above. But just as consumers spent more than expected then, at other times they have spent far less.

Thus all the statistical parameters (high income, high savings, low debt, etc.) may point to a jump in consumer buying, but consumers may still refuse to buy, forcing a seemingly logical forecast to fall flat on its face. An example of this occurred in early 1970. Forecasts for that year called for sales of 9.3 to 9.5 million automobiles. All the determinants of car buying backed this up. Yet because of some psychological quirk, buyers failed to spend, and sales actually turned out near the 8.5 million mark.

In a sense, this 1970 auto behavior seemed to represent a fundamental change in the attitude among a growing number of Americans toward their automobiles. Where they once viewed cars as fun and something special to own, they at that time seemed frustrated or bored with their cars. The novelty and the status of car ownership had disappeared. They looked at their auto as an appliance—something to get them economically from place to place and a thing to be replaced only when it wore out.

The implications of these more austere attitudes toward autos, if they should turn out to be long-range, could have far-reaching effects on Detroit. It could mean that Americans will spend less of their available income on cars. Indeed, some auto analysts are projecting future spending at the rate of 4.2 percent to 4.5 percent of available income—less than the average of nearly 5 percent since World War II. Such figures, of course, imply a slower rate of industry growth.

The more functional attitude also shows up in buyers' desires for smaller and more practical cars—and cars with less styling emphasis. The stress on size is somewhat threatening to Detroit, implying a reversal of the cycle toward bigger, more complex, and hence more profitable cars.

Another repercussion: As the interest in car styling fades, Detroit has begun to thin out its broad spectrum of models. *Automotive News,* a trade publication, figured there were only 341 new models in 1971—down 9 percent from the record 375 in 1970 and the sharpest cutback since World War II.

This reduced emphasis on new models and styling changes could also portend dramatic changes in the way cars will be sold. For one, it could knock into a cocked hat one of the industry's chief selling tools: the built-in obsolescence factor.

But all is not black as far as Detroit is concerned. New advertising pitches have already begun to replace the old ones. Thus you read more and more about safety, reduced pollution, and more economy per driving mile. To be sure, the long-term trend toward smaller cars probably means smaller average profit per car. But the de-emphasis on styling changes may prevent a total collapse of profit margins. Detroit will have to spend less per car on tooling, and more stable design will make it easier to automate assembly lines. Then, too, the growing availability of smaller cars could spur the trend toward two-car families.

The fickleness and changeability of the consumer is not, of course, limited to auto purchases. In a sense it extends across the board. This can best be appreciated by taking a look at Figure 8-1. It should destroy once and for all the belief that a perfect relationship exists between spending and income.

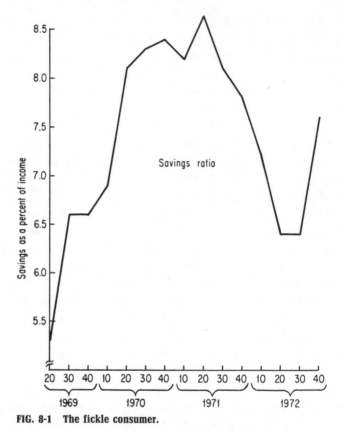

FIG. 8-1   The fickle consumer.

Thus between 1969 and 1972 people's savings inclinations varied all the way from 6 percent to up over 8 percent. Percentagewise that's a pretty big swing and the equivalent of close to $20 billion in GNP.

Conclusion: Taking the consumer for granted in the short run is fraught with danger. Maybe families will save about 6½ percent of their take-home pay over a 10-year period, but assuming they will spend this 6½ percent figure in any given quarter is clearly unwarranted.

DEBUNKING MYTHICAL ASSUMPTIONS

There are two sides to the "assumption" coin. So far, emphasis has been on the fact that numbers are highly dependent upon the validity of underlying assumptions. But it is equally important to keep in mind that assumptions themselves are often amenable to evaluation by judicious use of numbers. This, incidentally, is how the myth of a constant savings rate was debunked.

More to the point, statistical analyses, used correctly, can

"I'm sorry, Mr. Travis, but even here, at the Freedom University, seven times eight is fifty-six." (© 1969, *The New Yorker* Magazine.)

many times show up an unwarranted assumption for what it really is — pure fantasy. Thus anyone who still believes that open enrollment (the providing of a free college education for all those desiring one) is the answer to members of the lower class seeking upward socioeconomic mobility might do well to ponder the following:

A recent study shows that of the men entering college, some 20 percent drop out during the first year, another 20 percent drop out during the second year, and another 13 percent fail to graduate. In other words, more than half of those entering college failed to reach their objective: a college degree. These figures are much higher than those that prevailed in the early 1960s, when open enrollment was not a factor. (No wonder, then, such cartoons as the one appearing on page 125.)

The role of the open enrollees in swelling the dropout list is confirmed by the fact that dropouts tend to consist of the following types: those who (1) have not taken a college preparatory course in high school, (2) have a high school average of less than B, (3) come from low-income families, (4) come from a blue-collar-type family, and (5) have fathers with less than a college education. Nearly every open enrollee meets every one of these conditions.

In a sense, dropping out has also tended to make a bad situation even worse. That's because the open enrollees had hoped to use college as a means of upward mobility. But the fact that they completed one, two, or three years of college has tended to mean little, since the rewards of those who have only a partial college education aren't much better than for those with a simple high school diploma.

Nor are the rewards for graduating as great as is sometimes expected. That's because open enrollment and other factors are tending to lead to an oversupply of college graduates. In short, by seeking upward mobility for everyone, we may in the long run operate against such mobility. College degrees pay off well only so far as comparatively few hold such degrees.

In short, one needn't always take the current "conventional wisdom" at face value. Often a little judicious use of numbers and figures can debunk the assumptions on which they're based—and hopefully point the way to more meaningful solutions to some of today's pressing social problems.

This certainly has proved to be the case with respect to the false assumption that more money spent on education means better education (see Chapter 5). Few, for example, would any longer argue that the head start program to help low-income pre-kindergarten children has been an overwhelming success. Credit the change of heart to a numerical study of the results. Specifically, no matter what statistical yardstick one chooses to use, it is almost impossible to show significant improvement of ghetto children who attended a head start program over those who did not.

Nor need the statistical debunking of unwarranted assumptions be limited to the social sphere. A look at a few sobering figures showing the concentration of American industry, for example, would end once and for all the curious belief that ours is a competitive society. Thus three companies make nearly all the autos produced in the United States. And the top four companies account for over 90 percent of all the flat glass and electric light bulbs produced. And the list goes on and on. To argue competition in the face of all this borders on the ludicrous.

Then, too, there's the widely held belief that foreign trade takes away 3 million American jobs. The facts just don't fit this argument. For every job lost via imports, there's a job created on the export side. In short, a close look at the numbers shows that foreign trade is neutral rather than negative on balance—at least as far as job creation is concerned.

There's still another myth that statistics can easily debunk: namely, that most women work for pin money—and that a relatively high level of unemployment can be tolerated if it is mostly women rather than breadwinner men who are out of work. It can be shown, for example, that more than one worker in a family has resulted in a big difference between average worker and average family income. And if one takes time to analyze the difference, one finds that women workers are the major factor behind this change.

Thus over the last two decades some 16 million women have joined the labor force versus only 7 million men. And now women's earnings account for 25 percent of total United States personal income. Contributions of this magnitude can hardly be considered pin money—and those who produce it are hardly supplementary workers.

Indeed, over 6 million of the 52 million families in America in 1972 were headed by women. Moreover, it can be shown that 2.5 million husband-wife families at that time depended on the wife's earnings for over one-half the family income.

Need more documentation? Then consider the following: In many cases the married woman's income spelled the difference between poverty and middle-class comfort. Over 14 percent of all families in 1972 with only one breadwinner for example, fell below the poverty line, while fewer than 4 percent of all families where both husband and wife worked fell into the poverty category. This difference shows up in higher income categories, too. Thus 30 percent of the families where the wife worked enjoyed incomes of over $15,000 per year in 1972. On the other hand, fewer than 20 percent of the families where the wife didn't work achieved this income level. In short, the pin money theory of women working just doesn't hold up when subjected to the statistical facts of life.

Similarly, a look at pertinent numbers would quickly dispel such myths as that all blacks seek welfare handouts, that all people of Eastern European descent work in blue-collar jobs, and that Eastern bankers control this country's wealth. Unfortunately, such beliefs die hard. But their demise can be hastened by judicious use of relevant numbers and figures.

NINE

THE MISUNDERSTOOD AVERAGE

A man on the average is comfortable when he has one foot in boiling water and the other on a slab of ice.

Sounds like nonsense. And, indeed, it is. But this ever-popular barb helps spotlight one of the major misuses of averages. Specifically, an average is meaningless if the elements that make up the measure are greatly different.

Put another way, because each observation plays a part in the final result, the average can be distorted by extreme values — at least as far as the most well-known average, the arithmetic mean, is concerned.

EXAMPLE: A high-income family wants to move to a certain street — and decides to use the average income on the street as a yardstick for determining the advisability of the move. Further assume that there is one rich eccentric living on the block earning $1 million per year and 50 poor families also living there earning $5,000 each.

As such, the average income would be $1,250,000 divided by 51—or over $24,000 per family. On the basis of this simple calculation a decision is made to move to the street in question.

Obviously the decision would be a bad one for the high-income family about to move in because the typical family already on the street is poor —not nearly as well-to-do as the simple average would seem to indicate.

The family faced with this moving problem would have been better advised to choose a different average or measure of central tendency before making its final decision. Specifically, instead of using the popular arithmetic mean (adding up the values of all the observations and dividing by the number of observations), it might have been more useful to choose an average of position.

Looking at the above income distribution, the family thinking of moving might choose either the most frequently occurring or typical income (the mode) or the median income—the dollar income of the family in the middle of the income distribution. If either of these two choices were made, the average would turn out to be $5,000. In other words, both the modal and median incomes would be $5,000—a much more meaningful measure as far as the potential mover is concerned.

But there is a problem: the relative strangeness of these other measures of central tendency. Thus a recent survey by the author revealed that less than 5 out of 100 laymen regard the median or the mode as a true average. The reason isn't too hard to find. Since first grade, our schools have stressed the arithmetic mean, so that for most only this yardstick has become synonymous with "average."

Yet a little thought and observation would reveal that other measures of central tendency are intuitively used in many areas. If one sees a football team containing 10 relatively light men and one 350 pound giant, one would automatically tend to characterize the team as light, with one notable exception. Few would think of the team as heavy.

Also, there has been some increase in the use of these other measures, particularly where government statistics are involved. Thus, when the U.S. government publishes its average family income statistics, it is always in terms of median income. Many of the nation's so-called vital statistics are also measured in terms of this yardstick.

THE POPULARITY OF THE AVERAGE

The particular choice among the different types of central tendency is only one problem associated with averages. Equally important is the way people tend to view an average. Specifically, it can mean many things to many people: To some it's a handy way of quickly describing a mass of data. To others an average is the first step in a series of analytic statistical yardsticks. And to still others it's a convenient way of sweeping things under the rug.

But no matter how viewed, it is certainly high on the hit parade of quantitative statistical tools. The reasons aren't too hard to find. First and probably most important is the fact that everybody can grasp the concept of "average." Indeed, there is a natural propensity on the part of all of us to think in terms of central tendency. Ask almost anyone about the financial position of a doctor or factory worker, and the first thing that comes to mind is the average income of each.

And it makes good sense. We all tend to talk in generalities to avoid complex descriptions. And the average is tailor-made for this type of summation approach. It's a single number, it's precise, and, as noted above, it's understood by all. But equally important is the fact that it eliminates the need for qualifying statements. Thus an average income eliminates the need to talk about the range of salaries or the many differences that may exist among subgroups of workers within, say, the doctor or factory worker categories.

Another factor in the average's popularity is that it can be easily translated into handy rules of thumb. Take the Bureau of Internal Revenue. If your deductions are average, they'll go through without a hitch. But watch out if they're significantly above or below the calculated norm.

Your own firm also uses the average as a norm—particularly where expense accounts are concerned. Thus you will probably need little or no expense documentation as long as the outlays are in line with average experience.

That's not to say company brass won't approve a higher than average expense or that the IRS won't approve a big medical or charity deduction. But in both cases more extensive documentation will be required, because in both cases the figures vary significantly from the average.

To some extent even your own personal advancement depends on averages. If you are just average, you'll manage to squeak by. If you're above average, your chances of promotion are enhanced considerably. And if you're under average, you may be the first to get the ax when business turns sour.

Another big plus for the average: virtually everybody can compute it. In the case of the arithmetic mean, all one does is add up all the observations and divide by the number of observations. Even in the case of the less commonly used measures of central tendency, the formulas for the most part are relatively unsophisticated. Certainly no advanced degree in statistics or mathematics is required. And in almost all cases, they can be mastered by anybody with normal intelligence, a little formal training in the mathematical discipline, and the will to know.

Then, too, the average provides a starting point for further analysis. One has to start dissecting the mass of unwieldy data somewhere—so why not with the average, which in its own way can provide a rough, first-approximation definition of the variable being analyzed?

This starting point, however, should not also be a stopping point. For, as useful as an average is, it can provide only a limited amount of information. Sooner or later we will want to know something about the spread around the average. And if we are aiming to do sophisticated work—such as statistical sampling, projecting a trend, or analyzing the relationship between two variables—then the average has to be regarded as little more than a basic building block—the first step toward turning up the really meaningful information we need for intelligent decision making.

But perhaps the most cogent point for the average's use—at least as far as the orientation of this book is concerned—is the fact that it can hide almost as much as it reveals. For many whose main purpose is to deceive or distort, the average is a handy tool for keeping a lot of unwanted information out of sight.

HOW AVERAGES CAN DISTORT

Open up your newspaper, and if you look hard enough, you can probably find several examples of how averages are used as a smoke screen rather than as a vehicle for spotlighting the underlying truth.

The headline proclaims that GNP has soared at an annual rate of 8 percent in a given quarter. Unnoted—or fairly far down in the story—is the fact that price inflation accounted for more than half of the advance. Unemployment is proclaimed to have fallen—but again only short shrift is given to some of the really pertinent information that went into this average figure.

You are not told that in certain areas of the country unemployment may have gone up. Or, equally significant, that joblessness among minority groups and young unmarrieds is still a serious problem. Why bother the reader with all this? If he is satisfied with the statement that the unemployment average dropped from 5.6 percent to 5.4 percent, why shouldn't the communication media and the Administration releasing the figures also be satisfied?

The whole treatment of prices via indexes, which are essentially averages of many different prices, would fall into the same category. We read that the cost-of-living index soared during one period, and we are quick to blame labor and management for this runaway spiral. Nobody takes time to look into the figures and find out that bad weather (which pushed up food prices), rather than labor or management, is responsible.

And even if we take the trouble to look into the industrial segment of the index, we may be deceived—because this, too, is an average of many different items. Thus in the early 1970s, during one three-month period over half of the overall industrial price rise was due to shortages in fuels and lumber. So again the general tendency to put the blame on labor or management would seem to be out of order.

In all the above, one thing stands out. When an average changes, take the time to look into the specific components that may have been responsible for the change. Then, and only then, do we begin to get an understanding of what has been happening —and what we might do to correct an essentially unpleasant situation.

In still other cases, averages might be entirely inappropriate. Here's an example culled from the experience of firms trying to meet increasingly stringent antipollution curbs. At one time in the late 1960s, industry was using the "average" concept in its search for necessary corrective equipment. But a moment's thought should indicate the fallacy of this approach, for such averages would be completely unacceptable since they have led to mistakes. Specifically, their use would mean that permissible pollution limits could be exceeded much of the time. This, of course, would defeat the purpose of the equipment, namely, the elimination of a given percentage of a factory's pollutants at all times.

Another possible error: Many times people tend to forget that there should be some relation between an average and all of its individual values. Thus it can be misleading to lump together buying plans of, say, the upper and lower classes. Each class represents a distinct market, and each will undoubtedly spend a different portion of its income on probably an entirely different mix of goods. The grand average of both classes, then, can be quite misleading in determining sales strategies.

Similarly, the fact that a multidivision firm shows the same profit two years running can hide a lot of useful information. It may be that some divisions showed strong gains and others substantial losses. To accept the average without delving into the makeup of the components that went into that average ignores the very real possibility of improving the firm's position the following year—by either beefing up the weak sisters or putting more emphasis on the real money winners.

Here's another common pitfall: the tendency to project a change in averages under changing ground rules—a move which is not always warranted.

An example can perhaps best make this clear. If you can drive from A to B at an average speed of 30 miles per hour (when the highway speed limit is 40 miles per hour), how fast will you be able to travel on an average if the speed limit is lifted to 50 miles per hour?

The first inclination is to say the speed limit is being boosted 25 percent (10 divided by 40), so we can probably boost our average speed by the 25 percent—or up to 37½ miles per hour (125 percent of 30 miles per hour).

But this is not very likely to be the case. That's because it is not always possible to travel at the speed limit—and the lost time due to heavy traffic when the speed limit was 40 miles per hour is likely to be the same when the speed limit is upped to 50 miles per hour. In short, whatever the limit, you will probably be subject to the same unavoidable delays. Ergo, while some of the time you will, indeed, be traveling 25 percent faster, at other times you will be traveling no faster than before. Conclusion: the average-miles-per-hour gain will be something less than 25 percent.

Sometimes distortions creep in because the reporter chooses a type of average that will best bolster his position. Thus a Latin-American dicatator might choose the arithmetic mean to measure average income because he knows that his few rich cronies will pull up the average and hence put his nation in a more favorable light.

In any event, it's clear that this typical assymetrical distribution will yield different results for different measures of central tendency. Generally speaking, when a few high values are involved with many low values, the mode will yield the lowest result, the median the middle value, and the arithmetic mean the highest one.

Take the following list of numbers: 1, 2, 2, 2, 3, 3, 4, 4, 15. The arithmetic mean is 4 (36 divided by 9). The mode or most frequent value is 2, and the median or middle value is 3.

These three measures coincide only when the distribution of observations is perfectly symmetrical. But the chances of this happening in our uncertain world are quite slim—mainly because so many different variables influence a given measurement or observation. Thus if one were to take 100 people at random, it would indeed be surprising if their incomes, heights, weights, intelligence quotients, etc., were evenly distributed about the average.

In many cases it is probably best to calculate two or three types of average. If properly explained, this can reveal a lot more about the basic data than any one single measure of central tendency.

Then, too, averages of position like the median offer still further advantages that the arithmetic mean cannot hope to match. For example, it can be subdivided into smaller categories. Indeed, just as the median cuts the distribution into two equal sections, so quartiles can cut it into four equal parts; deciles, into 10 equal parts; and percentiles, into 100 equal parts. In short, by focusing on position rather than value, it is possible to make a more detailed study of any distribution.

The most common use of these smaller subdivisions is in the field of education, where a student may be said to be, say, in the upper decile, or 10 percent, of his class. But there is nothing that disqualifies this type of analysis for use with consumer-oriented problems.

Suppose, for example, that we are interested in the price/ earnings ratio of stocks. By using this approach, we could say, for example, that 1 percent, 10 percent, or 25 percent of all corporations have P/E ratios above a certain level. This, in turn, could be of considerable use to a stock market analyst appraising the relative merits of specific stocks.

THE TRICKY BUSINESS OF ANALYZING RATES OF CHANGE

The arithmetic mean, median, and mode by no means exhaust the possible measures of central tendency. Indeed, one other, the *geometric mean,* is quite important for specialized applications. Thus in cases where we are attempting to average rates of change rather than absolute numbers, the geometric mean is the only correct approach.

This can probably best be understood by the use of a simple example. Double the price of a product the first year from $1 to $2, and triple the price the second year from the already existing $2 up to $6. Question: What was the average rate of rise over the two-year period?

Your intuitive answer would probably be 2½ times—the average of a doubling and a tripling over the two-year period. Sounds cut and dried—and relatively straightforward.

But let's stop and think about it for a moment. If the average rate of price advance was indeed 2½ times each year, then we might logically expect that two consecutive 2½-times-a-year advances would bring the final price level up to $6.

Does it, though? After the first year the price would be $2.50. After the second year, and another doubling, the price would be $6.25.

But this hardly jibes with the $6 price we arrived at when the problem was originally proposed. Ergo the paradox: Use of the average suggests a $6.25 price. Use of the actual doubling and then tripling yields a $6 price. The average, then, is obviously incorrect—or somewhat high, to be more precise. But why? Simply because the average we used (the arithmetic mean) was the wrong one. As suggested at the beginning of this section, the geometric mean would have been the correct approach, since the rule states that this latter measure of central tendency must be used to average rates of change.

How then do we compute the geometric mean to arrive at the true rate of change? It is defined as the nth root of the product of n observations. More specifically, in our simple example it means multiplying the doubling (2) and the tripling (3)—and then taking the square root of the product.

This means $\sqrt{2 \times 3} = \sqrt{6} = 2.45$. For those who want to check this out, multiply the geometric average rate by itself, and the answer is quite close to 6.

This is a mighty valuable formula for all types of growth problems. Thus, if you know the beginning and end figures (in our example, $1 and $6), it is relatively easy to find out the true rate of growth. In other words, we know prices went up six-fold over two years, but don't know the true rate of growth. Again our geometric mean formula comes to the rescue—yielding 2.45 times each year as the correct answer.

Nor need the geometric mean be limited to observations over two periods. You might want to assess the growth rate over 10, 50, or even 100 years. The same basic formula would apply— but the arithmetic would become more difficult (in the case of 100 years, you would be taking the hundredth root of the product of 100 numbers).

Fortunately, this arithmetic has been tabled for us—with the answers contained in the basic compound interest rate tables found in almost any business arithmetic book.

These tables which embody the principle of the geometric

mean can also be used to solve other related growth problems, including (1) the estimation of some future level (assuming a constant rate of growth) and (2) the number of periods that would be needed to achieve a given level at a given rate of growth.

The importance of the geometric mean in the above contexts can't be overemphasized for the solution of many of today's business, economic, and demographic problems. All these areas are basically concerned with growth rates—and when you're talking average in this context, you're talking geometric mean.

The curious might raise an interesting question at this point. In the above example the arithmetic mean yielded an answer somewhat on the high side. Does this always occur, and why?

The answer is, Yes. Whenever the arithmetic mean is erroneously used to average rates of change, it will yield an excessively high answer. As for the why: The hint was first given in Chapter 2, where ratio charts were discussed. The geometric mean is analogous to the ratio chart—giving equal distance or weight to equal rates of change. On the other hand, the arithmetic mean is analogous to the arithmetic-grid chart, giving equal distance or weight to equal absolute changes.

As an example—a doubling from 1 to 2 is given the same weight as a doubling from 2 to 4 when the geometric mean is used. But when the arithmetic mean is used, the "2 to 4" jump (2 units) gets a bigger weight—thereby leading to an overestimation of the average rate.

AVERAGING RATIOS: ANOTHER
COMMON PITFALL

Many times when data are expressed in ratio form—cents per pound, number per dollar, or miles per hour—the arithmetic mean will yield incorrect results. If there's any doubt on this score, try solving the following problem:

Assume a man travels from one point to another at 20 miles

per hour, and then travels back to his original starting point at 60 miles per hour. The question: What was his average speed during the round trip?

Chances are nine out of ten people will come up with an answer of 40 miles per hour—with their thinking running something along these lines: The man went at 20 miles per hour and he came back at 60 miles per hour, so obviously his average speed is 40 miles per hour.

Here again a seeming paradox arises if you think about the problem a little more carefully. Specifically, is it not also true that it took him 3 minutes to go 1 mile and only 1 minute to come back? If this is true, then obviously, it took him only 2 minutes on the average to go a mile. But if one takes 2 minutes to go 1 mile, in 60 minutes (one hour) one will obviously travel 30 miles. That's the same thing as saying 30 miles per hour— yet our arithmetic mean calculation suggested a 40-miles-per-hour rate.

Obviously, both can't be right. But which one is the correct one? In this case the second calculation gets the nod—because it can readily be seen that he traveled 3 minutes per mile going and 1 minute per mile coming back—for an average of 2 minutes per mile or 30 miles per hour.

What we have done in this second calculation is to use the so-called *harmonic mean.* This is defined as the reciprocal of the arithmetic mean of the reciprocals of the individual items. In other words, 1 divided by the average of $\frac{1}{20}$ and $\frac{1}{60}$. Try this calculation, and you'll end up with 30 miles per hour, the correct answer.

How do we know when to use the harmonic mean and when to use the arithmetic mean? The answer lies in the rule of thumb: If the ratios being averaged have the same denominator, use the arithmetic mean. If they have the same value in the numerator, use the harmonic mean.

Let's examine our original problem again. The problem was

posed in terms of miles per hour (20 miles per hour and 60 miles per hour). Since the miles traversed would be the same coming and going (i.e., the numerators), it follows that the harmonic mean is the one to use.

There are many other types of ratio problems where the choice between arithmetic mean and harmonic mean must be made. Take a simple shopping problem. If bananas in one store sell for 50 cents a pound and in another for 25 cents a pound, what then is the average price if you buy 1 pound in each store?

Here, too, let's be careful. Again our first inclination is to use the arithmetic mean—averaging 25 cents per pound and 50 cents per pound—and coming up with an average price of 37½ cents per pound.

But are we right? Let's use our rule of thumb. The problem is expressed in cents per pound (15 cents per pound and 30 cents per pound). The conditions of the problem state that we buy 1 pound in each store. In other words, the denominators of the ratios are the same, so the arithmetic mean is correct this time around. Now let's see what would have happened if we had expressed the problem somewhat differently. Suppose we said we can buy 4 pounds of bananas for $1 in one store and 2 pounds of bananas in another store for $1. What is the average price per pound of bananas?

Using the arithmetic mean yields 3 pounds of bananas for $1 on the average—or 33⅓ cents per pound—obviously the wrong answer. Why? Because the same 1 pound is being purchased in each store. That is, the numerators of the ratios are remaining the same, thus calling for the harmonic mean treatment, which would yield the correct result of 37½ cents per pound.

If, on the other hand, the problem was posed in the same way but the question asked was: What would be the average price if $1 was spent in each store? Obviously the denominators (dollars) would remain the same, thus calling for arithmetic-mean treatment and an average price of 33⅓ cents per pound.

If further proof of this is needed, think of it in terms of purchasing 2 pounds of bananas for $1 and 4 pounds of bananas for an additional dollar. That's 6 pounds of bananas for $2 or 3 pounds of bananas for $1—or 33⅓ cents per pound.

Try these problems on your friends. Chances are few will ever stumble on the correct answers—and even fewer on the correct approach; for how many have ever heard of the harmonic mean? And yet it is a valid measure of central tendency—and often an indispensable one when the averaging of two or more ratios is involved.

AVERAGING PERCENTAGES AND AVERAGING AVERAGES

One of the most common mistakes made by laymen involves the averaging of two or more percentages. Thus, if goods prices go up 5 percent and service costs 10 percent, the first inclination is to say that prices went up an average of 7½ percent.

This would be true only if outlays for goods and services were the same. But if consumers spent more on goods than on services, the average increase would be closer to 5 percent. On the other hand, if they spent more on services, the average advance would be closer to 10 percent.

The implication is clear: Each percentage must be "weighted" by its relative importance.

In a similar vein, if one market surveyor finds that 8 people out of 100 percent prefer brand A and a second survey finds that 10 out of 50 (20 percent) prefer brand A, we would calculate the average preference for brand A by weighting the individual percentages by their appropriate sample sizes. Thus:

$$\frac{(8\% \times 100) + (20\% \times 50)}{150} = 12\%$$

There are times, however, when it may be best to ignore the different sample sizes, and instead use a different set of weights. In short, rules are made to be broken when common sense dictates.

Take the example of the student who takes two examinations during a semester. He receives 70 percent on the first test (based on 50 examination questions) and 90 percent on the second test (based on 100 questions). If we weigh each percentage by the number of questions, we arrive at this term mark:

$$\frac{(70\% \times 50) + (90\% \times 100)}{150} = 83\frac{1}{3}\%$$

But is this realistic? Not really, if you stop to consider that each test represents a level of accomplishment for half a term's work. Keeping the latter point in mind, the tests might better be weighted equally, yielding an average term mark of only 80 percent.

The moral is clear: When the importance of each percentage is dependent on some factor other than the number of items included in that percentage, the percentage must be weighted with this other factor.

The possibility of wrong weighting also crops up when averaging averages. Again, one would normally weight by the number of observations in each average. But this, too, can lead to incorrect conclusions under certain conditions.

A case in point: You are thinking about buying an economy car which can yield relatively high gasoline mileage. An auto salesman quotes a test which shows that *(a)* 20 of his cars driven in city traffic yielded 15 miles per gallon and *(b)* 40 of his cars driven only on high-speed highways resulted in a 20-miles-per-gallon average. Question: What mileage can you expect?

It would be erroneous to calculate your answer by weighting

each sample average by the number of cars in each test, because the observed overall average mileáge is based on a certain ratio of city-to-highway driving—a ratio which you might not necessarily adhere to. Let's assume, for example, that you plan on utilizing your car three-fourths of the time on the highway. Then your true average mileage might be calculated by weighting 15 miles per gallon by one-fourth and 20 miles per gallon by three-fourths. Thus:

$$15 \times \tfrac{1}{4} + 20 \times \tfrac{3}{4} = 18\tfrac{3}{4}$$

By following this procedure you are again weighting by the key factor that determines miles per gallon: not the number of cars in each sample, but rather the type of driving you expect to be doing.

THE AVERAGE IS NOT ENOUGH

Even in cases where the appropriate measures (weighted correctly) are used, the average can result in misleading conclusions. Much of the problem lies in the fact that central tendency is only one, albeit an important, aspect of data measurement. In many cases, the spread about the average is of equal significance. Clearly a store where prices range all the way from bargain-basement to luxury levels is a lot different from one which emphasizes medium-priced merchandise. Yet the average price in both stores may be about the same.

In short, any accurate description of a mass of data must include the degree of variability as well as the degree of central tendency. Weather conditions are another case in point. Two areas may have the same average rainfall over the year. In one case it all falls during a three-month rainy season. In the other case the precipitation is evenly spread throughout the year. For agriculture and a variety of other purposes, these two types of climate could hardly be described as the same. Yet that might well be the case if only averages were used.

Study of the range about an average also gives us the opportunity to look into the causes of variability. In the case of the rainfall example, the spread is, of course, due to the fact that the two areas under study are, geographically speaking, quite different. But there are usually two other general dispersion-causing factors at work, too: the heterogeneity of the data and inherent variability.

The heterogeneous factor crops up whenever two different populations are being lumped together. Thus the income, racial makeup, or even politics of a metropolitan area will show a lot more variation than an affluent suburb, because the former is made up of many different groups of people and the latter of one—usually homogeneous—group.

Inherent variability is another kind of problem altogether. Toss 10 coins several times, and chances are the number of heads and tails you wind up with will differ each time. Indeed, the odds of getting 5 heads and 5 tails when tossing the 10 coins (something you might intuitively expect) are quite small. This chance factor was discussed in Chapter 7.

Little can be done about controlling this kind of chance variation, but that's not to say that it can't be used as a powerful analytic tool for decision making. In many instances, for example, the amount of such variability can aid in choosing between alternative procedures or processes.

Say, for example, you're buying a TV set. Past experience has shown that on the average brand A lasts longer than brand B, but that the chances of winding up with a lemon are also better with this same brand A. Do you opt for the set with the longer average life (brand A) or the set that is more likely to last over its normal life span (brand B)? You pay your money and take your choice. But at least by knowing the average life and inherent variation, you are making a conscious decision—something that would not be possible if the variability of TV sets were not known.

Manufacturers are also faced with this kind of problem. Should they play up longevity or reliability? Again their decision would depend on the type of product being sold. Obviously, flashlights or a tire will sell better on a reliability pitch. On the other hand, low-priced autos (where economy is the main consideration) will sell better on the longevity argument. Volkswagen and Volvo, for example, have used this longevity approach with great success in recent years.

A few words on how you might actually compute variability may also be in order. The generally accepted technique is by the use of the so-called *standard deviation*—an approach which essentially calculates the average variation from the arithmetic mean.

Knowledge of the standard deviation can provide needed information in another way, too. When there is a normal amount of variation, the standard deviation tells us the percent of cases that might fall within a given portion of the range. Statisticians, for example, claim that when a distribution has a "normal" amount of variation, about two-thirds of all observations will fall between plus or minus 1 standard deviation of the mean. Plus or minus 2 standard deviations would cover 95 percent of the cases, and plus or minus 3 standards deviations would cover virtually every case (see Figure 9-1).

Thus if an examination resulted in an average mark of 70 and the standard deviation was 10, we could then expect that about two-thirds of the marks would fall in the 60 percent to 80 percent range.

This average-plus-standard deviation approach is indispensable for "marking on a curve." Specifically, a certain number of standard deviations above a given average determine the cutoff points for both A and B grades.

This approach, however, can lead to inequities. If an instruc-

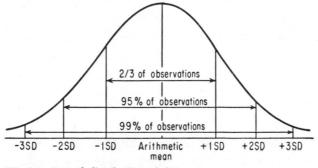

FIG. 9-1   Normal distribution of data.

tor has a class full of bright students, many of them are being penalized by the underlying assumption that there are poor, fair, and good students in every class. Similarly under the "curve" approach, a class of poor students will tend to fare somewhat better than they ordinarily would.

Other problems sometimes arise when two series are being compared with respect to their variation. Generally speaking, two such series cannot be compared for variation when (1) their means are significantly different or (2) they are expressed in different units of measurement.

The latter is a relatively easy problem to solve. Divide the respective standard deviations by their averages, and you put both on a comparable basis. This can probably best be understood in terms of a simple example.

If you're comparing variation in income with variation in height, then obviously the standard deviations alone are not enough—for one would be in terms of dollars and the other in terms of inches. But if you divide each by the appropriate average, the problem disappears. By dividing dollars by dollars and inches by inches you wind up with a relative yardstick, independent of the units of measure.

A relative measure is also needed where the appropriate variations, even though expressed in the same units, vary significantly. Thus the dollar variation in an affluent section where incomes vary, say, between $30,000 and $50,000 a year will be far above the dollar variation in a poverty pocket where incomes vary between only $3,000 and $5,000. Common sense would tell us that the expected variation about their respective averages is about the same. And this is indeed the case when the variations of $20,000 and $2,000 are divided by their respective average incomes.

In most cases, the calculation is done in terms of standard deviation rather than ranges, for this adds another element of standardization to this measure, which has come to be known as the *coefficient of variation*.

At other times, relative statistical measures can be developed by dividing by the standard deviation. This is usually the case when individual observations are compared with norms. Assume this time around that you are applying for employment and are asked to take intelligence and aptitude tests, with your chances of getting the position depending on how well you do on each test.

Let's further assume you score 120 and 180 on intelligence and aptitude tests where the norms or averages are 100 and 150, respectively. Since variation about these norms is different for each of the tests, we obviously can't compare the "20 above average" on the intelligence test with "30 above average" on the aptitude examination. But we can reduce these two differences to comparable magnitudes by dividing them by their respective variations (standard deviations).

Following through on this, and assuming standard deviations of 10 and 15, we can then compare the observed differences by dividing by their standard deviations. The results would be 20/10 and 30/15, or 2 and 2, respectively.

In the parlance of statisticians and social scientists, you would have achieved a standard score of 2 on each of the tests— well above average.

Or, put another way, your deviation above the norm in both tests was the equivalent of 2 standard deviations. But remember that 2 standard deviations on either side of the average cover 95 percent of all observations. As such, if you accept the validity of such tests, they suggest you are in the top few percentiles of the population when it comes to both intelligence and aptitude—and chances are, you'll get the job.

TEN

THE SEMANTIC TRAP

A rose is a rose is a rose.

Maybe so. But a number is not always a number—at least, it doesn't always measure the phenomenon it purports to measure.

Depending on how one may choose to define one's terms, (1) inflation has been either accelerating or decelerating, (2) much or little headway is being made on the desegregation front, (3) the unemployment problem is being licked or is growing worse, and (4) profits are either rising or falling. The list could go on and on, for arguments on both sides can be convincing, provided one picks the "right" set of statistics.

In short, one man's definition of any of these terms is not always the same as the next man's. And if the definition in question is different (either by choice or by design), it is often possible to come up with a diametrically opposed conclusion.

The desegregation issue provides a classic case of how one can fall into this kind of definitional dilemma. Thus in 1970 the federal government stated that 94 percent of the Deep South public schools systems had been officially desegregated. But upon further examination it was found that official integration was often accompanied by de facto segregation.

In short, one's estimate of progress in desegregation depended on one's definition of the word "desegregation." Among the many indirect methods used by the South at that time to circumvent official desegregation were (1) the administering of ability and achievement tests to determine class makeups, a technique which resulted in virtually all-black or all-white classes in some schools; (2) the trek to private schools, which left desegregated schools predominantly black; (3) the gerrymandering of school districts to keep the races apart; and (4) allowing students to choose their classes (they nearly always opted to stay with their own kind).

To be sure, a few Southern school districts actually were desegregating at the time. Indeed, even the most pessimistic of civil rights leaders admitted to some progress. What they objected to, though, was the fact that the listing of the 94 percent desegregation figure was misleading—that it was convincing many well-meaning people that the integration job in the Deep South was nearing completion.

At other times misunderstandings stem from differences in basic perspectives. Many of these revolve around the old "Is the bottle half full or half empty?" dilemma.

A simple example best illustrates this problem. In 1968 steel buyers built up a substantial 13 million-ton inventory hedge in anticipation of a strike which never materialized. Once a new pact was negotiated, the problem was to dispose of the excess metal as quickly as possible. After three months, buyers succeeded in just about halving their previous hedge.

Those who were satisfied with this stock-paring pace accentuated the positive—namely, that more than 6 million tons had been whetted off inventory totals in an extremely short period of time. Other analysts, however, preferred to point out that despite paring progress, some 6 million tons of excess metal was still hanging over the market—enough to keep mill operating rates at very low levels.

People's understanding of words can also differ as a result of sharply different economic perspectives. Thus our annual 3 percent to 5 percent rise in the price level over the past decade seems inflationary to us, but appears pretty much stability to some Latin-American country where annual price rises in excess of 50 percent are not uncommon. In the same vein, a $5,000-per-year income may be regarded as horrendously low in the United States but the height of opulence in some of the less developed nations of the world.

But even where geography and economics are kept constant, the list of possible "definitional" misunderstandings defy counting. Below is just a partial sampling, from areas of critical importance to both consumers and businessmen.

WAGE INCREASES AND THEIR IMPACT

Some of the sharpest disagreements in recent years have centered around the question of measuring both the magnitude of pay boosts and their impact on costs and prices. Because of widespread disagreements on just what constitutes a pay hike and a cost increase, the same union contract could be evaluated as either inflationary, neutral, or even deflationary.

As to the pay hike itself, differences usually arise because one man's definition may include fringes (social security, unemployment, life insurance, profit sharing, etc.) while the next man's may not. Make no mistake about it, inclusion or exclusion can make a whale of a difference. That's because there has been increasing emphasis on fringes in recent years—resulting in a faster rate of growth than might be suggested by looking at dollar-and-cents pay rates alone. Upshot: Widely publicized hourly pay increases tend to understate the true percentage increase in overall labor costs.

There are plenty of good solid facts to back up this burgeoning fringe effect. According to one recent tabulation, close to 30 percent of United States industry's total labor costs were being attributed to fringe benefits in the early 1970s. That's nearly double the percentage reported two decades earlier. So one ignores such fringe benefits at one's own risk.

And risk there is. Not only does exclusion of such benefits tend to underestimate the annual rate of labor cost increases, but it also tends to give us misleading estimates of real purchasing power. Unions like to calculate purchasing power by dividing hourly wage payments by the price level—conveniently forgetting to add in fringes. But these fringes contribute to a worker's standard of living every bit as much as a dollars-and-cents hourly wage boost.

That's why in the late 1960s union claims that inflation had wiped out all of their wage gains were suspect. If the unions had taken the trouble to add in fringes, the total increase would have come out slightly higher than the price increase over the corresponding period. Conclusion: Rather than falling, true or real purchasing power, or the workers' standard of living, had actually risen slightly over the period in question.

Then there's the overtime complication. Wage rates are computed on the basis of both regular pay and overtime plus regular pay. And in some instances the differences can be significant. In 1971, for example, the regular hourly pay for a blue-collar worker in the United States averaged out at near $3.44 per hour. Add in overtime, and the figure grew to around $3.57 per hour.

This may not seem like too much of a difference on an hourly basis. But if you multiply it by the 40 hours worked per week and the 52 weeks per year, it adds up to a healthy sum. Thus the 13 cents per hour differential multiplied by 40 hours and 52 weeks yields a not insignificant $270 per year. Remember, too, this is for all manufacturing. In some highly seasonal industries the gap runs upwards of $500 per year.

The difference becomes especially important at contract negotiation time. Unions prefer to stress their lower regular pay, while management usually opts for the higher overtime concept; ergo, you often hear two different sets of figures being bandied about — one suggesting the need for big catch-up raises, the other for more union moderation.

Still another wage pay complication arises from the confusion over wage rates and unit labor costs. The two are not the same. Indeed, they're quite different — and again one can signal inflationary pressures while the other may be quite encouraging as far as wage-cost pressure on prices is concerned. Generally speaking, wage costs tend to overestimate the impact of a labor cost boost because they don't take into effect any offsetting productivity advance.

The point is that labor costs to any firm are not simply the money-plus-fringes that it has to lay out, but rather these outlays in relation to worker efficiency. If costs go up 5 percent and productivity goes up 3 percent, the net additional per-unit cost to the company is only 2 percent. And it is this latter cost that is pertinent to the company in terms of cost pressures and ultimately the asking price of the finished product.

The following simple illustration can perhaps best illustrate the difference between wage costs and unit labor costs. Assume you are producing a product with labor constituting the only input. Your labor force can produce 4 units per hour at a pay scale of $4 per hour. So obviously the unit costs of production are $1 per hour. If you then sell this product for $1.50 per unit, you are making 50 cents per unit profit — or $2 in total profit.

Now let's see what happens a year later. New equipment has permitted a 50 percent increase in productivity. In other words, each worker, instead of turning out 4 units per hour is now turning out 6 units per hour. But labor, aware of what has happened, demands and gets an equal 50 percent increase in pay—to $6 per hour.

Let's then look at our new set of cost calculations. The unit cost of production (again assuming that labor is the only input factor) again comes out to $1 per unit. And if we continue to assume $1.50 selling price, the profit per unit is still 50 cents. But this time around, we are turning out 6 units, so the total profit has increased to $3 (where before it was only $2).

Summing up then, a 50 percent increase in productivity, coupled with a 50 percent increase in labor costs, has also permitted the company to rack up a 50 percent increase in profits. This before-and-after situation is summarized in Figure 10-1.

The above should help dispel one of the popular myths nurtured by management—namely, that if wages increase at the same rate as productivity, there is nothing left for management. The fact is that there is always something left—with the exact

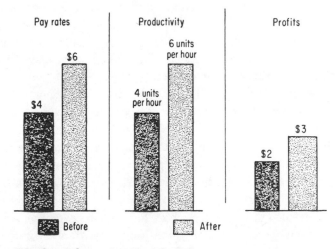

FIG. 10-1   Labor costs and productivity.

percent determined basically by the labor content of the product being manufactured.

The myth, luckily, seems to be on the way out. Certainly, the government has finally recognized the truth. Witness the wage-price guidelines which have been formulated over recent years. They state that wage increases should be in line with productivity gains, and that if such a relationship is maintained, then prices will remain stable and profits will tend to grow in line with historic rates.

The above distinction between wages and unit labor costs should dispel still another popular myth—the one that states that high American labor costs have priced our goods out of international markets. The same basic argument as above can be applied here. Sure, American wage costs are well above those prevailing in other parts of the world. But so is our productivity. As such, our unit labor costs, on average, are probably no higher than those of any other industrial countries.

Some tentative studies have tended to support this thesis. But even if you are unwilling to accept such figures, common sense might lead you to the same conclusion. Specifically, if our costs were so much out of line—as many claim—how can you explain the fact that our trade (exports less imports) has been in rough balance over the last decade? This would hardly have been the case if our costs were twice and more above some of our leading competitors', as many claim.

To be sure, these critics can point to products where our unit labor costs are higher—electronic components and photographic equipment are two good examples. On the other hand, it is equally true that we have an advantage in sophisticated machinery—witness our consistent positive trade balance in these lines over the past decade. In short, on a unit cost basis (and that's the only basis that counts) we have an advantage in certain areas and our competitors have an advantage in other areas. But that's what makes a ball game. More important, economists tell us that if everybody produces what they can make more efficiently, then everybody's standard of living tends to rise. Economists have a name for this, calling it the *law of comparative advantage.*

THE MYTHICAL COST-OF-LIVING INDEX

Everybody talks about the cost of living these days, but in most cases not everybody is quite sure of just what they're talking about. The typical layman tends to equate the rise in the cost of living with the rise in the government's consumer price index. Unfortunately, the two are not the same. Specifically, the government index generally has tended to understate the true rise in living costs—and that's one reason why so many of us find it hard to make ends meet—even though the consumer price index is rising at only a moderate rate.

The confusion between the two concepts stems from the fact that the consumer price index measures only the changes in prices of a fixed market basket of goods. The cost of living, on the other hand, in addition to taking into account the price rise in such a fixed market basket of goods, also figures in our rising expectations. The fact that we now want and expect better homes, finer clothing, and longer and more luxurious vacations has, in essence, meant an upgrading of the fixed market basket of goods—and hence has made the rise in the cost of living greater than that indicated by the rise in consumer price indexes.

The following anecdote can perhaps best illustrate the confusion between price and cost of living. Consider a storekeeper who complains about inflation, citing the doubling of his electric bill over the past year. When it is pointed out to him that utility rates haven't changed over this period, his classic reply is, "No, but I just put in central air conditioning."

The problem here, of course, is that the storekeeper has failed to differentiate between a price rise and a standard of living advance. The sharp rise in utility costs to which he refers pertains to his own decision to raise his own cost or standard of living—not to any noticeable rise in the price level. But the government doesn't have a true cost-of-living index; it has a price index. And the two can vary substantially, particularly when standards of living are changing rapidly.

Lest there be any doubt on this score, take your salary of 10 years ago—and then inflate it to take care of the actual rise in prices over this period. Chances are you would be unable to live on this inflated figure today—despite the fact that you managed quite well on that sum a decade ago. The problem, of course, is that your standard of living has probably jumped enormously over this long period of time.

In any event, it is important to remember that price indices are designed to measure change in a constant market basket of goods rather than upward shifts in living costs. Factors other than prices that affect the cost of living are best measured by government family-budget studies.

THERE'S MORE THAN ONE PRICE

Our prolific statisticians have developed so many inflation indexes and subindexes that it is often difficult to come up with the one "correct" gauge of price movements.

The actual yardsticks and what each supposedly measures are discussed in detail in Chapter 11. More important at this point, however, is the need to recognize that many different yardsticks exist, and that, depending on the one chosen, it is possible to either play up or down the inflation story.

Take 1971 experience. The consumer price index that year rose 4.3 percent—reflecting, in large part, a big rise in food and service prices. Those who were fearful of inflationary pressures cited this measure to justify the need for corrective action. On the other hand, the wholesale price index rose at a much more modest 3 percent rate that year—and was widely quoted by the administration in power to prove that it had the inflationary spiral under control.

Who was right? Actually, both sides were, because both indexes measure price changes—but from different vantage points. If one were thinking in terms of family living expenses, then the consumer price index would certainly be the more pertinent measure of inflation. But if one were thinking in terms of the impact of inflation on business, then the wholesale price index would be the more useful one.

Again the lesson is the same: There are no absolute wrongs or rights where definitional differences are involved. Rather, one must take pains to find out what measure is being used and whether that measure is pertinent to the problem being analyzed.

At still other times the unsuspecting consumer is purposely misled as to the price in question. Take the case of a store running a so-called "sale"—when, in fact, no such "sale" price is being offered. The unscrupulous storekeeper might plaster his windows with signs offering 30 percent off list, thus implying

(© 1965, *The New Yorker* Magazine.)

some eye-opening bargains. But in many instances such discounts are normal practice. So the consumer winds up paying the regular market price.

Sellers are still able to get away with this maneuver as long as they don't use the word "sale," which, in effect, promises a lower-than-normal price. By using "discount off list" they are not lying in the narrow sense of the word. So there's little that the Federal Trade Commission and other consumer protection agencies can do about it.

Other price interpretation problems (again stemming from the wealth of pricing intelligence) crop up when comparisons between indexes are involved. Thus distortions can arise when one compares a price on one level (say, wholesale) with the price of the same product on another level (say, retail). There are similar difficulties when one compares a list price with a market or "transaction" quotation. More about these comparison-type pitfalls in Chapter 13.

DIFFERING BUSINESS AND ECONOMIC DEFINITIONS

Anyone who has ever tried to do serious work on such key aspects of the business cycle as retail sales, profits, consumer incomes, inventories, or even foreign trade is well aware of the definitional pitfalls that await him. Here it's not so much that the originator of the statistics in question wants to deceive; rather, the problem lies in the fact that one man's definition of what constitutes any of the above may simply not dovetail with the next man's. A few examples should suffice to illustrate the difficulties involved:

1. *Measuring consumer income.* Even such a simple concept as family earnings is open to many different interpretations. The three most common yardsticks are: gross income (what your salary calls for), your take-home pay (what you get in your pay check), and real income (what your pay check will actually buy). For most consumers it is the third measure that is crucial.

But there are other confusions as well, stemming from the fact that there is now usually more than one jobholder in the typical family. Ergo, further distinction has to be made between individual income and family income.

In short, the man of the house may still be the main breadwinner, but proportionately he is bringing home far less of the bread than he used to. This was pointed out above in Chapter 8 where the pin money myth of women's labor was exploded. But it shows up even more vividly in the disparity between the growth in average weekly wages paid to individuals and the growth in total family incomes.

In the 1965–1970 period, for example, the weekly wage average climbed some 25 percent—a substantial gain itself. But the median family income rose a whopping 42 percent. Conclusion: Gauging the rise in the American family's buying power by following weekly wage rates alone may have been valid at one time, but not any more. The analyst who follows such a course may be greatly underrating the purchasing-power growth of the typical American family.

2. *Defining profit.* Earnings performance can be equally ambiguous. Thus profits can be going up, down, or sideways, depending on what type of yardstick one is referring to. There are a number of possibilities. There's pre-tax or after-tax dollar income, pre-tax or after-tax earnings per dollar of sale, and pretax or after-tax earnings per dollar of stockholder's equity.

Even such concepts of dollar earnings are not without their ambiguities. Do they refer to gross earnings, net earnings, operating revenues, or total revenues? The differences involved are enormous, but all four are often referred to only as earnings, leaving the hard-pressed reader to guess at what precise concept the analyst has in mind.

To be sure, each of the above dollar magnitudes is valid —and has its place in our financial accounting system. But the trouble arises when these measures are used interchangeably, with the actual choice at any given time depending on what specific impression one wants to leave with the reader or listener.

The choice among the various profit margin concepts—since they are basically percentages—was discussed in greater detail in Chapter 3.

3. *What is a retail sale?* This isn't as easy to define as it may seem. Some people think of it in terms of overall consumer outlays (including services), while others equate the concept only with purchase from a regular retail outlet. The two are not the same. In fact, on occasion one of these yardsticks may be falling while the other is rising. And even when they do move in the same direction, the magnitude of change is almost never the same.

Again, both measures are useful. A retailer, for example, would be most interested in merchandise sales from bona fide consumer outlets—and hence would be more likely to use the retail data. But an economist, anxious to gauge GNP trends, would find the consumer outlay concept a more useful one.

Retailers also have another option—the use of department store sales, which is a still narrower measure of consumer spending. Here the choice of yardstick would depend on the type of operation the analyst was involved in. Certainly the trends wouldn't be the same. The increasing movement of department stores to the suburbs, for example, has made for a relatively bigger growth rate for these corporate giants—often at the expense of the smaller, nondepartmentalized stores.

4. *Inventories—Level versus change.* Another kind of problem complicates the task of measuring business stocks. The underlying difficulty here stems from the basic confusion over the concepts of level and change. Thus one set of government figures zeros in on the level of inventories each month—the billions of dollars of such stocks in the hands of manufacturers, wholesalers, and retailers. This is certainly a must for gauging the health of the economy—particularly the development of any inventory imbalances.

But confusing the issue is another set of figures (this time issued quarterly) which puts the spotlight only on inventory change. This, too, is important, but from a different angle. Only by knowing the change can we properly distinguish that part of any GNP advance which is attributable to final demand and that part which is due only to inventory buildup.

Try to reconcile the two sets of figures, and you get even more confused. That's because the data coverage is somewhat different and because each of the two series is valued on a different basis (for conceptual reasons). Thus the monthly series is put in terms of book value (in line with prevailing accounting practices), while the GNP series is expressed in current value —because GNP, by definition, is the current value of all goods and services produced. The moral: Don't blindly accept any government series. Check the concepts behind it—and then see if these concepts dovetail with the purposes of your own study.

5. *International balance or imbalance.* Just as inventories are subject to definitional differences, so are balance of payments statistics. In one recent year, depending on the concept used, this critical measure of our international financial health was either badly in the red or moderately positive. Compounding the confusion was the fact that the same statistical agency (the Department of Commerce) was issuing both versions. Both versions were, of course, tenable. But not so the position of commentators and armchair economists who chose one or the other version—not so much on the basis of conviction or need, but rather on the basis of whether they wanted to paint a rosy or a gloomy picture.

Examples, culled from many other areas of statistics, could also be presented, pointing up similar misuses based on definitional differences. And in a sense there's little the recipient can do about it, except to insist that the definitional concept be given whenever more than one set of numbers purporting to measure the same phenomenon is available.

ELEVEN

PITFALLS OF AN INDEX

We are fast becoming a nation of index watchers.

Whether it be the so-called consumer price index (sometimes erroneously referred to as the "cost-of-living" index—see Chapter 10), the Dow Jones Industrial Average, or even the farm parity ratio, the scenario is always the same: we wait with baited breath for the next daily, weekly, or monthly report—to dissect, analyze, and pontificate over the latest minor index change.

If one number had to be picked out as the public favorite, however, it would have to be the consumer price index (CPI). And with good reason; for this is the yardstick that best gauges consumer or worker purchasing power. What wage earner, for example, hasn't compared his latest wage increase with the rate of consumer price advance, with his level of satisfaction or dissatisfaction roughly proportional to the gap between these two magnitudes?

But while use of this measure has become virtually universal now, few (except for the professional statistician or economist) have ever taken time out to get a better understanding of this yardstick—more specifically, of its pluses and minuses and what it can and cannot do.

Perhaps many more should, for they would find that it is not nearly the perfect measure of inflation it purports to be. Among the many charges leveled at this widely monitored yardstick are (1) failure to reflect fully the quality improvements that have taken place; (2) failure to add new items and to subtract old items from the index rapidly enough; (3) a built-in overstatement of prices because the index tends to weight heavily items that are of shrinking importance in the consumer's budget; and (4) delay in reflecting the effect of new methods of distribution (discount houses, etc.).

These and other inherent weaknesses will be discussed in greater detail below. But first a digression on what makes indexes tick. One thing for sure, it's more than of academic interest—for only by knowing how and why indexes work is it possible to pinpoint and anticipate their shortcomings.

First and foremost, indexes do fulfill a needed function—that of a guide to composite changes in a large number of different commodities. It's a vital function, too—for in most types of policy decisions, the crucial question is not so much how an individual price has changed, but rather how the average price of many different items has changed.

How does one go about building up such an average? For one thing, it's not quite as simple as it seems. How, for example, do you average the price of wheat and rice? Obviously the former is far more important, so it must be given more weight. But how much more?

This might best be explained by examining how a typical index such as the CPI is constructed. This can probably best be described in a series of simple steps:

1. Determine a normal or base period on which to calculate all subsequent index changes. (More about how the base is decided below.)

2. Next take a large sampling of consumers to pinpoint their spending habits during this base period. From this you are then able to deduce the products that John Q. Public, the average consumer, has purchased in the base period. In the parlance of statisticians, this is the "base-period market basket of goods."

3. These items are then divided into subgroups — such as food, durable goods, and services. A fixed weight is assigned to each of these categories on the basis of their relative dollar importance during the base period. Thus if the typical consumer earmarked 30 percent of his budget for food, then food prices would get a 30 percent weight when all the prices were being averaged. Similarly, each of the subgroups can be built up from individual products, with the weights of such individual products again being derived from the base-period market basket.

4. Each month send surveyors out into the field to monitor the current prices of the goods and services contained in the base-period market basket.

5. Calculate precisely how much it would now cost to buy the base-period market basket.

6. Take this current cost of the base-period market basket and divide it by what it had cost in the base period.

7. If a basket that now costs $6,000 had cost $5,000 in the base period, we would then wind up with a figure of 1.2 (6,000/5,000). By multiplying this ratio by 100 (to get it into percent form) we end up with an index figure of 120. Translation: Prices today cost 120 percent of those prevailing in the base period. Or, put even more simply: Prices have gone up 20 percent since the base period.

The big pluses of the index approach are:

- It simplifies the price averaging of many different types of products. It's the one case where "apples and oranges" can be combined to yield a meaningful figure.
- It facilitates temporal comparisons. Just looking at the reported index number gives you a "quick and dirty" percent change from the base period—the accepted point of reference.
- It can be used equally well for averaging physical and price changes. Indeed, when it comes to averaging the output of different products such as autos and TV sets, index numbers offer the only viable solution, since each item can be weighted by its appropriate base-year value.

But an index's strength is also part of its weakness. When everything is thrown into the hopper, distortions inevitably arise. To be sure, they're not serious enough to negate the whole approach, since a fairly good approximation is certainly better than a wild guess or no figure at all. Nevertheless, in some cases such distortions can modify the overall meaning of an index change. And for that reason alone, a detailed study of index-number limitations would seem to be in order.

THE PROBLEM OF AVERAGING

Since an index is essentially an average, all of the difficulties associated with averages pertain also to indexes. The basic problem here is that the change in the index will not always reflect the changes felt by each individual who uses this average.

Housing is a case in point. At any given time, the cost changes faced by apartment dwellers and home owners may be quite different. Thus at a time when mortgage costs are rising (a major element of home-owning costs), rental housing may well be quite stable. Thus, depending on your own particular situation, housing costs may be rising or stable. In short, the "average" cost in this instance means very little.

Again in the housing sphere, consider our senior citizens, many of whom tend to live in rent-controlled apartments. They are only marginally affected whenever rental costs rise. Again, their costs are not average costs.

In a similar vein, an older home owner who purchased his home in the early 1960s probably financed it with a low 5 percent mortgage. A home owner who bought in 1970 may have paid 8 percent for the same basic loan. In other words, when mortgage rates rose to 8 percent and pushed up housing costs, it had little effect on the older home owner who had nailed down his lower-cost mortgage years before.

Finally, an illustration from the transportation area. Back in early 1972 the cost of living in New York City took a sharp jump when subway fares soared 40 percent (from 25 cents a ride to 35 cents). But this hardly even ruffled the feathers of the more affluent New Yorker, who continued to use his car to get to work.

WEIGHTING WOES

As noted above, the various items that make up an index are not always of equal importance. Ergo, one has to "weigh" the various elements that enter into an index to reflect these differences. While the concept of weighting is simple enough, the execution can become quite complicated. Indeed, a change in the type of weighting used can often result in a changed index reading.

Most index makers choose one of three approaches: base-year weights, given-year weights, or combination weights. And it does make a difference, for what may be important in one year may be of lesser importance in another year. Looking at each of these approaches separately:

The base-year approach was used in the example above describing how the consumer price index (CPI) was computed. Indeed, this is probably the most often used technique, with a majority of government indexes using this weighting system.

Recall that in the CPI illustration we assigned weights equivalent to their basic importance at the time the market basket was surveyed (the base period). If the weights are kept constant at base-period levels, any change in the index over time must, by definition, be attributed only to price.

But as popular as the base-period weight approach is, it is not without its shortcomings. The big problem is that the base-period market basket of goods tends to change when price changes. In other words, the CPI, as constructed, ignores the tendency on the part of the public to substitute relatively low-priced goods for relatively high-priced ones. Take meat over the past decade or so. Families have been spending a larger part of their income for chicken than for beef. And with good reason: chicken has stayed relatively stable while beef tags have soared.

Expressing all this another way, price changes lead to consumer substitution of chicken for beef. But because we use a fixed set of base-period weights, the index is unable to pick up this shift in buying patterns. Don't underestimate this "substitution" effect either, for it can add up over a period of years. Specifically, by giving too much weight to products with rapidly rising prices, our CPI probably tends to overstate the rate of inflation.

By how much? It's hard to put a precise figure on this upward bias. But according to a recent estimate by economists of the First National City Bank of New York, the overestimation could be in the order of ½ percent a year. In short, after five years or so, the "true" price rise could be in the neighborhood of 2½ percent less than the figures suggested by the reported CPI.

(Incidentally, analysts suspect that the consumer index may also be inflated by rising hospital costs. That's because, while these costs are undoubtedly rising, it's not always the consumer who's footing the bill. More and more it's the employer who's picking up the tab. To the extent this is happening, the rise should be earmarked for the business rather than the consumer sector of the economy.)

It might be argued that because of this substitution effect it might be better to use given-year weights. This proposal would have index makers use the weights in the current year as a basis for gauging importance. But this approach has the opposite type of bias effect. It gives undue weight to those commodities that have been stable or have been experiencing relatively low price advances. The reasoning is essentially the same as given above for base-year weights—except that the positions as between the rapidly rising prices (say, beef) and the relatively stable ones (say, chicken) have been reversed.

There are still further problems for those who would opt for this second (given-year weights) approach. It takes time and money to calculate a changing market basket of goods. A survey would have to be done at least once a year—making the cost of index construction prohibitively high and forcing long delays in publication (calculation of such weights would lag the availability of prices by many months).

There would be some conceptual difficulties, too. If weights change every year, how does one compare the current year with the previous year? Obviously, one can't, because both prices and weights are changing, making it impossible for the recipient of the final figure to separate out the crucial price component.

Other statisticians suggest the use of combination weights—those using calculations based on an average of the base-and given-year weights. And there are a few government indexes that do use this approach. But the three criticisms which apply to given-year weights—costliness, publication delays, and lack of year-to-year comparability—apply equally as well here.

Finally there are those who would junk weights entirely and just take a simple or unweighted average of the prices involved.

In most cases, of course, this is unsatisfactory—primarily because different prices do have different importance. And in most cases these differences are substantial.

But, like everything else, there are exceptions to the rule that weighting be applied to indexes. One might be the special case where the weights are approximately equal. Another might be where such differences in importance are irrelevant to the purpose to which the index is to be put.

Uncle Sam, for example, puts out a weekly "spot" index of raw-material prices which is unweighted. The rationale: The index does not purport to measure changes in true price level but rather "grass roots" changes in supply and demand. An advance suggests that business (not price) may be improving on the primary level, while a decline might point to a future decline in general business sentiment. In short, because the index is used as a barometer more of business trends than of actual price level, the weighting (which would be necessary for price-level changes) can be dispensed with.

Studies have been made on how these different weight options can change index readings. In one, involving the consumer price index, results indicated a 2 percent difference in the index level after five years. That's not inconsiderable—especially where wage rates are escalated in accordance with movements of this index.

COVERAGE GAPS

There is always the problem of which materials or items to include. While it is seldom necessary or even wise to include each and every item or transaction, the actual decision must depend in large part on the individual situation. Consider the problem of the government attempting to work up a consumer price index. It is physically impossible to include the literally hundreds of thousands of items available in our consumer-oriented society.

The only way out of the box is to determine a typical market basket—and then assume that the items not included in the market basket move in line with those included. Of course, this assumption isn't always warranted—and this is one area where errors can certainly creep in.

Sometimes sampling can help. Thus the general practice of Uncle Sam's statisticians is to steer clear of outright enumeration of all the items available on the market. For one, it's just too expensive; and secondly, time required for such an approach would result in an index publication lag of several months. Instead, the government has compromised—enumerating all major expense components, such as autos and beef, and taking a random sample of all others. This enumeration-sampling combination also carries through to the type of sellers (e.g., supermarket versus corner grocery store), with the big outlets more adequately covered.

In short, the government indexes make sure they capture price changes in all important areas, and then sample the rest in the hope that their sampling results will yield a good approximation in these other less important areas.

From the above discussion it follows that most indices are based on a sample of commodities which have been purposely selected rather than chosen by random methods. Therefore, the standard statistical techniques for evaluating the error in a sample are not applicable.

However, experience over a long period of time suggests that an index becomes increasingly reliable as the group of prices becomes larger. As the Bureau of Labor Statistics (BLS) puts it: "The reliability of a subgroup is greater than that of a product class, a group is more reliable than a subgroup and the all-commodities index is more reliable than a group index."

SPECIFICATION CHANGES

Nor is it enough to decide on the number and type of products to be included. If men's shirts are to be included, the question then arises: What type of shirt, what material, what style, and what price line? Obviously, one can't choose every single variation. So a choice must be made.

That the consumer price index always measures the most relevant items also is in some doubt. The only "men's business shirts" priced were all-white and 100 percent cotton until 1966, when blends of synthetic fabrics started qualifying. But not until five years later did officials decide that a shirt could be priced if it had a "long-point collar" and colored stripes.

But relevant or not, specifications cannot be changed willy-nilly, for a cardinal rule of any price index is that the only variable that should effect the change in the index level is price. Therefore, it is highly desirable for prices used in the index to adhere to rigid specifications insofar as this is possible. Note that this applies to more than just physical characteristics. Thus specifications should include quantity discounts, credit terms, delivery charges, modes of packaging, package sizes, etc.

Problems arise when any one of these specifications changes. Take our man's shirt. Assume we started pricing a 50-50 cotton-polyester blend. After a few years, producers might decide a 40-60 blend is more advantageous. The product is now different, and so are its costs. How do we deal with this problem? There are adjustments that are possible, but they are all approximated—and hence open up the possibility of introducing error.

Changes in nonphysical characteristics, however, cause the biggest headaches, because you can't "see" the differences. And if a data collector inadvertently forgets to "adjust," that entire segment of the index can be thrown out of kilter.

PRODUCT MIX CHANGES

Difficulties often arise because the proportion of different products purchased just won't stay put. Actually this can arise from two different factors:

Consumption patterns and tastes change. What might be a

typical market basket of goods purchased by the average wage earner 10 years ago is not a typical one today. Air conditioning was a luxury 10 years ago. Today, for many, it has become a necessity. Put another way, the weight or relative importance of individual items making up an index change over time—and sooner or later the index must be adjusted to take these changes into account.

Make no mistake about it, these changes can be quite significant. Thus a 1969 study of Miami and Portland, Oregon, consumption patterns showed that the importance of food to total consumer spending in those cities had fallen by 15 percent since the weights were last determined, and that recreation outlays had more than doubled from the roughly 4 percent weight assigned a decade earlier.

The timing of new-product entries into the index also has an impact. Innovative items are generally brought to the market at high initial prices and, as production and competition increase, the prices begin to fall. The price behavior of ballpoint pens and television sets in the 1940s is a good example. Early inclusion of new items would clearly give the consumer price index a downward bias at a later date. On the other hand, later insertion would exclude the favorable impact of falling prices.

But there's just so much that can be done. It can be shown that any market basket of goods—no matter how many attempts to adjust it are made over time—will gradually become less and less representative of the true world. This is true even when insertions and deletions are made every year, for such insertions and deletions are essentially a "patch up" rather than a major adjustment of underlying consumption patterns.

This "patchwork" approach also shows up in the way these products are inserted and deleted. There is no exact way to measure the impact on a price index, for example, of a new product replacing an old one. You know today's new price when it was introduced. But how does it compare with two and three years ago when only the price of the older commodity was available?

The statistician can come up with only a rough estimate known as "linking." This is essentially a method which assumes that at the point of shift the old price and the new one are comparable on an index basis. Put another way, the assumption is that any price difference that does exist when the new product is entering the index and the old one leaving the index is due to a difference in quality. This is hardly a sophisticated technique; and when the two products involved are subject to different types of market forces, it could well lead to questionable results. But as index makers say, there's precious little else they can do short of a wholesale revision of the index every few years, something that is hardly feasible.

THE DISTRIBUTION LAG

Closely allied with the new-product problem is what to do with new types of outlets. Thus in recent years there has been a trend toward "bargain" retail stores—particularly food chains, food supermarkets, and discount houses. While the lower prices in these stores are reflected in the index, they may not be given full weight because of their sudden growth.

Uncle Sam's statisticians have made a crude estimate of this omission, at least as it applies to the food component. Over the period from December 1955 to December 1961, the reported food component index rose by 8 percent, while an index adjusted for the increased proportion of sales by food chains increased by 7.3 percent. In short, the failure to monitor the increased proportion of sales made by chains resulted in an unwarranted 0.7 percent increase in the food price index.

A QUESTION OF QUALITY

The automobile turned out today is not the same product turned out a decade ago. There are new safety and antipollution devices on cars, tires wear longer, and the vehicle supposedly is easier to handle. All these changes created problems, because a price index theoretically is designed to measure price changes of constant-quality goods.

If quality deteriorates at constant price, for example, the consumer is obviously getting less for his money, and an index should rightfully consider this a price rise. Similarly, if one gets a better product at constant price, he is getting more for his dollar—or, in effect, a price reduction.

The following chapter details how quality adjustments—as well as other types of adjustments to index numbers—are made.

But for all the pains the government takes, some quality improvements slip by unnoticed. One BLS official recently said, "It is likely that the incomplete representation of quality improvements imparts some upward bias (makes the index higher than it should be) in the long run."

However, keep in mind that quality can be a two-way street. For example, there seems to have been an evident decline in the quality of some services. Anyone who has attempted to have his car or TV set repaired can easily attest to that. There is also reason to think that in a period of intense prosperity, sellers tend to substitute higher price lines for low-end items at prices which reflect something more than the improvement of quality.

Then, too, some analysts say the index makers may too blindly accept manufacturers' claims of quality improvement, adjust away actual price increases, and thus understate the true climb in costs faced by consumers.

Actually the most serious problem is making quality determinations for services. Thus it's almost impossible to measure a longer wait in the doctor's office versus the fact that you're apt to get more scientific and better treatment. Consequently indexes make few quality adjustments in the service area. "We need to do a lot more work in services," observes a top government official.

Generally speaking, every time wages or fees in some sector of the service industry rise, the index shows a rise in service prices. The entire increase is then attributed to inflation, and no account is taken of possible productivity increases and quality improvement. While in some cases the assumption of zero increases in productivity may be true, there are few patients who would want to be treated by a doctor who was still relying on nineteenth-century techniques, even if his fee were substantially lower.

TRANSACTION VERSUS LIST PRICE

More often than not, published or reported prices are nominal or list prices serving merely as a basis for price negotiation. Nor can one rely on a constant adjustment in such prices to correct for what is usually an overstated price. The actual prices at which transactions are made will vary from time to time, depending on prevailing economic conditions, supply and demand factors, and the negotiating ability of the parties involved in the transaction.

Why are list prices unrealistic? Two factors probably contribute. First, sellers may be fearful of stiffer government controls, and want to get hikes on their books before Uncle Sam moves. More important, there are always "special deals" going on. But because of the possibility of federal charges of price discrimination, sellers are generally loath to report such reductions to government number collectors.

It follows, then—theoretically at least—that only transaction prices should be used. But this is more easily said than done. For one thing, buyers and sellers often prefer to keep their deals private. But more important—even if they were willing to disclose the correct price—the task of collecting the appropriate data would be enormous. Consider, for example, a steel price. It is a lot easier to go to a few sellers in this oligopolistic industry than to the literally thousands of buyers who might consume the given steel product.

The government from time to time has attempted to do some

research or pilot studies on transaction prices, using buyer information. Thus a few years ago it shifted its aluminum ingot quotation from a list to a transaction or market basis. The result: The former 29 cents-per-pound list quote was dropped to a 21 cents-per-pound market level. In other words, there was a difference of 8 cents per pound, or 27 percent, between these two prices. To be sure, aluminum is an extreme case (and this was one reason why Uncle Sam's number collectors chose this particular commodity for a test case). But it serves to illustrate the need to know just what kind of prices an index represents.

Bolstered by the above findings, the government in recent years has attempted to shift more of its quotes onto a transaction basis. But cost and manpower would seem to preclude any all-out shift to this more realistic pricing approach.

To a large extent, then, subjective judgment is the only way out of this impasse. When a buyers' market and price shading are known to exist, one can probably assume with reasonable certainty that official price indexes overstate the existing price situation. On the other hand, in times of shortage and rising prices it is usually safe to assume that a list-oriented price index is fairly accurate.

BASE-PERIOD SHIFTS

There's always the question of choosing a base period against which to compare subsequent price changes. If a recession period (for example, 1970) were picked as a base, the index would have an inflationary tone, since subsequent prices were on the uptrend. Similarly, a deflationary trend might be built into an index where the base period occurred at the peak of the business cycle. Sometimes it is preferable to use the average of two or three years as a base; many government indices, for example, have generally been based on the latter type of approach.

Changes in the base are also musts as the gap between the base period and the current period grows. Certainly, when index numbers of 200 and 300 are reported, it becomes a lot more difficult to relate them to a base of 100. The point is that 100 has become a convenient base, and that when numbers greatly exceed that base, they become hard to work with, unwieldy, and, more important, less meaningful to the average reader.

Then, too, there's the product mix problem alluded to above. Usually when a base is changed, there is a complete overhaul of the index—a time when products can be reweighted, new products added, and old ones thrown out. The problem of quality adjustments also would seem to dictate an occasional base-period update, for there are subtle and not-so-subtle quality "errors" which accumulate over a long period of time. On the other hand, chances of such errors are much reduced when, say, today's car is compared with the car turned out last year or the year before last.

Last, but not least, the reliability of the average represented by the index becomes more and more questionable as time goes on. It is possible, for example, that the individual commodities being averaged will show wider and wider dispersion. This makes averaging—the raison d'être of index-number use—somewhat less meaningful. The average price of meat when pork and beef are fluctuating within 1 or 2 percentage points is a lot more meaningful for policy decisions than when 10 or 20 percentage points separate the individual products.

No wonder, then, that governments periodically update their index bases. Generally speaking, such moves are made every 10 years. Thus, if you had been following United States government indexes over the past 30 years, you would have had to deal with three base periods: 1947 to 1949, then 1957 to 1959, and 1967. These changes usually are followed by private index makers, too—because only then is it possible to optimize one of the major advantages of index numbers: the facilitating of comparisons among many different series. Thus, if you were a wage earner, you would want your union to put its wage index on a 1967 base—if only to see how wages were moving in comparison with consumer prices (a government 1967-based index).

THE "RATE-LEVEL" BOOBY TRAP

The typical price index is designed to measure change, not absolute levels of prices. This can best be appreciated by considering the consumer price index for two different metropolitan areas. Specifically, these figures don't really show how prices in one area compare with those in other areas. Thus New York may have a higher price index than, say, Chicago, but that doesn't necessarily mean that prices are higher in New York than in Chicago. All it means is that prices have risen faster in New York than in Chicago since the base period.

Similarly, one can't compare cost-of-living indexes for two different countries. Prices (and hence price indexes) in most Latin-American countries, for example, have risen at a far faster clip than they have here in the United States. Ergo, their index readings are higher. Yet it would be the height of folly to suggest that the price level is higher in these Latin-American countries.

It's worth noting, too, that this level-rate of change distortion isn't necessarily limited to index numbers. Thus an unsuspecting layman reading that a ditch-digger received a 50 percent wage boost might be tempted to infer that ditch-diggers were "getting away with murder." If he stopped to think a moment and realized that the raise was from a very low $2-per-hour level, he might have some second thoughts about his original conclusion.

The basic misunderstanding here, as well as in index numbers, is the same—confusing a rate of change with an absolute level. There is no a priori reason to assume that the two are in any way related.

CHOOSING THE RIGHT INDEX

Too much of anything can create problems. And the same can be said about price indexes. There are just so many available— from private as well as government sources—that it oftens becomes difficult to know which one to use.

One thing for sure, they're not interchangeable. Thus wholesale prices can be falling while consumer prices are rising. And rarely does the GNP price level—the price average of all transactions in the economy—ever coincide with either of the two. There's good reason for all of this, for each of these yardsticks measures a different segment of the economy, and hence it is designed to serve a different purpose.

Probably the most familiar is the consumer index—for this is the one that purportedly affects all our lives by influencing the purchasing power of the dollar. But even here care must be taken—for while it measures prices of more than 400 individual goods and services, it is supposed to represent only the typical market basket of a city wage earner or clerical worker. These groups represent only about 40 percent of the total population. The index market basket, for example, wouldn't be typical of a farm family, and so the Agriculture Department has worked up a special index for the rural consumer.

Then, too, what about the growing number of senior citizens? The spending patterns of these oldsters by no stretch of the imagination could be assumed to be even remotely related to that of a young-married family. The food these senior citizens buy, the clothes they wear, their drug outlays—even the way they spend their leisure time—would all seem to dictate a special index.

Despite the above facts which make the CPI a questionable yardstick for the entire economy, it is widely used for a variety of purposes, aside from measuring the inflationary impact on consumer purchasing power. For millions of workers it determines how much extra they will get in their pay checks via cost-of-living escalator clauses—clauses which automatically guarantee workers a stipulated cents-per-hour wage advance when the CPI moves up by a given amount.

Similar use of the CPI is reported for determining the magnitude of pension payments. And there has been some experimentation—applying cost-of-living escalation even to bond payments and annuities.

But all of this could be self-defeating. At least that's the claim of some economists, who label such clauses an "engine of inflation." Their thesis: Prices rise; so wages rise via escalation. This wage escalation, in turn, could set off a secondary round of price increases, which means more wage escalation, etc.

The argument seems exaggerated, for both wages and prices always move up with a lag. But, nevertheless, like everything else, there are more than a few grains of truth in it—and it should be considered when setting national wage-price priorities and goals.

All this talk about the CPI often makes people forget about the wholesale price index (WPI). In many ways this is even more crucial for gauging inflation. For the WPI measures changes at the primary level, and hence provides useful clues to what will be happening on subsequent levels in future months.

A case in point: In early 1973 the wholesale price of food soared. Economists, noting this, quickly predicted that the relative slowdown in consumer prices might soon be over. And they were unfortunately right. By the second and third quarters of that year, consumer prices were rising at a 5 percent to 6 percent annual rate—twice the pace noted when wholesale food prices first began to move up the price ladder.

Later that year steel prices moved up at wholesale—leading to the prediction that thousands of steel-containing consumer items might soon be boosted. And so they were. So once again the WPI proved a useful barometer of future movements in the more widely followed CPI.

As might be expected, then, the WPI is closely monitored by businessmen and economists as a clue to future inflationary movements. But again there are other important uses. One is for contract escalation for long-lead-time purchases. Under such clauses if raw materials go up in price, the seller is entitled to an automatic price rise. Then, too, businessmen find the WPI much more useful than the CPI for gauging the purchasing power of the business dollar.

Finally a few words may be in order on the use of the GNP price deflator, which includes wholesale prices and retail prices, as well as a lot of other elements that enter into a nation's overall pricing structure. Perhaps its biggest use is to convert the dollar GNP advance into real growth. Specifically, divide dollar GNP by the GNP deflator, and you wind up with a useful measure of a nation's increase in physical output. (More about this in the next chapter, where this "deflating" process is discussed in greater detail.)

But, this GNP price deflator can be biased. Generally speaking, most economists feel that, because it fails to take into account government worker productivity, it tends to be somewhat on the high side, thus resulting in an upward bias. Other criticisms of this yardstick are: (1) it comes out quarterly—not nearly often enough to be of use to economists trying to monitor the economy on a day-to-day basis; and (2) there is a lag of nearly two months. Thus the first-quarter estimate is usually not out until nearly two months after the quarter is over—not nearly early enough for timely policy decisions.

But with all these indexes and subindexes, seldom is the published yardstick tailor-made for the use that it's put to. Thus even the consumer price index is not a true measure of the general purchasing power of the dollar. It does not include prices of securities, real estate, or a host of other things a consumer can purchase.

And the index's treatment of taxes has spurred growing complaints. Increases in sales taxes, excise taxes, and real estate taxes on homes do show up in the price index, but federal, state, and local personal income taxes don't. To properly "price" the immense income tax burden, officials argue that the government would have to decide if a tax-rate increase is offset by an increase in the quantity or quality of governmental services. "If it were easy to do such things, we would have done so long ago," observes one Washington statistician.

Even some segments of the index are open to question. For example, there's a running feud between government index makers and auto insurers. Actual collision repair costs have jumped 93 percent in the past decade, the industry asserts. This spurt, say insurers, makes their premium-rate increases seem like gouging, because the CPI showed only a 39 percent climb in auto repair costs. Government men retort that the repairs covered in the CPI are properly limited to slower-rising maintenance-type items (such as tune-ups), and that most faster-increasing collision repair costs should not be included because they are paid by the insurance companies, not by consumers.

In any case, the user has to compromise—and realize that he is using only an approximation. That's not to say, of course, that one can't improve on an index—either by building a brand-new one or by making adjustments in an available yardstick.

On the latter score, government agencies like the Bureau of Labor Statistics (BLS) may be induced to develop a variant of a particular price index for, say, contract escalation. Or, if this isn't feasible, the private index maker, by using the basic government data, can do the recalculation on his own.

TWELVE

A MATTER OF ADJUSTMENT

"Nothing ever really fits the way it should." How many times have you heard this complaint? Indeed, how many times have you used it yourself?

Whether it be the suit or dress just purchased, your child's new toy supposedly all ready for assembly, or even the rug cut to cover your 24×15 foot living room, the problem always seems to be the same: seldom can we use something without an alteration or adjustment.

And so it goes with numbers, too. The raw figures gathered by the experts—even if accurate, reliable, and free of distortion— are seldom ready for final application.

Actually, two types of number adjustments are usually in order. One is aimed at fitting the data to one's own particular needs. Some of the difficulties involved here were touched upon in the previous chapter, where the possibility of "tailoring" index numbers to a given problem was discussed. Little more can be said about this type of adjustment because there are almost as many possibilities as there are problems.

But there's another type of adjustment that's quite amenable to generalization—the kind which is meaningful to almost all users of numbers and figures. Thus, if you're concerned about inflation, you want to know more than just the dollar amount of your pay check. More important is the question: What does it mean in terms of purchasing power? Actually the adjustment in this case is a fairly simple one: present income in deflated terms—the statistician's special term for taking the price influence out of the data in question.

There are other types of "common" adjustments, too—including those used to take the seasonal influence out of numbers and those used to take the quality factor out of price.

It is important to distinguish all these general types of adjustments because many different series can be presented in several different "adjusted" forms. Unless one is aware of the differences—and of what particular series is being presented at a given time—serious errors can result.

Consider the politician seeking reelection who points out that sales rose 10 percent in a certain July to December period. Sounds great. But what he has failed to point out is a lot more important: mainly, that sales might normally be expected to rise this 10 percent from the summer low to the Christmas peak. Ergo, the increase isn't really that outstanding. Indeed, it's downright disappointing, for it suggests no real growth. It is for this reason, of course, that seasonally adjusted data (numbers with the seasonal effect filtered out) have become almost a necessity for meaningful business analysis. To be sure, some people feel that this seasonal adjusting has been overdone, and that in many cases it is wholly irrelevant. Hence such anecdotes as the following:

A young lady employee comes to the chairman of the board and asks whether she can wear a pants suit. The chairman then proceeds to call his senior officers together. Their decision: The lady in question may wear pants suits. Some months later another young lady approaches the chairman and asks for permission to wear hot pants. Another meeting is called—but this time the decision is negative. But then the company statistician intervenes, pointing out that the executive decision is inconsistent—since it is apparent that hot pants are nothing more than pants suits seasonally adjusted.

But putting levity aside for the moment, no reputable statistician would deny the need to differentiate among the many possible ways a number or figure can be presented to a listener or reader—for only then can the correct version be distinguished from the incorrect ones. It is to this task, then, that the current chapter is addressed.

DEFLATION

We all deflate—although we are not always conscious of it. The worker who gripes that his 6 percent wage boost went up in smoke because of a similar rise in consumer prices is implicitly performing a "deflation" operation. When you get right down to it, he is doing nothing more than dividing his bigger pay check by higher prices and finding out that his "real" pay, or pay in terms of purchasing power, has not gone up at all.

And this is essentially the definition of *deflation:* dividing dollar magnitudes by price to come up with an estimate of physical or volume magnitudes. Government statisticians do it all the time when they talk about real growth. In essence they are (1) taking the change in reported or dollar GNP, (2) dividing it by the GNP price index, and (3) coming up with real or deflated GNP—another name for real growth.

And for many purposes this is probably the most important figure to watch when trying to assess the progress of an economy; for the standard of living can improve only if the physical output of goods and services increases. If the reported dollar increase is all due to price, then obviously we are no better off when it comes to sharing the physical fruits of our labor.

Ignoring the inflation impact in this case can lead to serious overestimation of our economic progress. Thus in the 1970 to 1972 period, more than one-third of the GNP advance was due to price. That's a pretty hefty number of dollars when you consider the close-to-a-trillion-dollar economy that prevailed during that period.

But coming closer to home, wage increases, to be meaningful, must also be viewed in deflated terms. Ditto company sales — particularly if the figures are to be used for future capital spending estimates. The same might be said about inventories — for if your stock levels are up only because of higher prices, you might suddenly find that these higher figures are inadequate to meet the higher production schedules projected over the next few months.

In short, when analyzing dollar data, there is an urgent need to deflate figures to arrive at real change.

Happily, the statistical techniques for accomplishing this are basically simple. It can usually be done by using a calculator or computer, relatively quickly and easily — provided, of course, that one has a reasonably good idea of how fast or how slowly prices are rising. On the latter score, there is plenty of price information available — both from private sources and from Uncle Sam. The government, for example, has consumer and wholesale price indexes made up of thousands of prices — presented separately and in various combinations. In other words, chances are there's a pretty good price index available to the consumer if he wants to filter the price effect out of a particular series being analyzed.

How does one actually physically accomplish deflation? Obviously, the concept is clear enough: Divide value by price to get physical volume or the actual number of units. But in the real world we are more often than not dealing with averages of many products expressed as indexes. Thus we may have the true value of, say, sales of our firm, but no equivalent dollar price. On the latter score we would probably have only a price index which approximates the average price received from the sales of our goods.

Question: Do we get any meaningful figure when we divide a true dollar magnitude such as sales by a price index? Fortunately, the answer is, Yes. We end up with sales in "constant dollars."

EXAMPLE: If sales rose from $100 million in year x to $200 million in year y at a time when our price index rose 25 percent (from 100 to 125), what was our physical increase in sales?

We proceed by dividing the dollar sales figures ($100 million and $200 million) by their respective price indexes (100 and 125). The results: $100 million and $160 million. Interpretation: Sales in year y went up $60 million in terms of the price level prevailing in year x. That's just another way of saying the physical volume of sales went up 60 percent. In short, by dividing dollars by a price index, we have effectively deflated our figures.

Such a deflation is a must when two series are being compared, where one is on a volume basis and the other on a dollar basis. Assume for a moment you're selling a certain type of small appliance and want to estimate company sales on the basis of consumer income. If the physical number of units sold is being used, then income must be deflated, for you must compare two variables on either a volume or a value concept. Use one "value" variable (money income) and one "volume" variable (unit sales), and you end up with a meaningless mishmash.

In other words, be consistent. Use values or volume throughout. Some people like to work with both types of relationships, one based on a volume concept, the other on a value concept. Strange as it may seem, the relationships can sometimes be quite different, thereby resulting in somewhat different projections.

But coming back to our hypothetical problem, unit sales require deflated income. So we would probably divide the consumer-income dollar series by a measure of price changes in consumer income, or, more specifically, the consumer price index—thereby getting consumer income with the price effect factored out.

There's another "plus" for deflation—particularly when the price of the product being sold is extremely volatile, rising or falling by substantial percentages each year. Attempting to forecast dollar sales, given the above conditions, is extremely risky, for such a forecast would entail projection of price as well as volume—increasing substantially the probability of projection errors.

If the forecast is to be used for future capital expansion needs, why expose yourself to needless risks—especially when the volume data are likely to be the more readily available type anyhow?

Volume forecasts are usually more accurate for another reason, too: The factors which determine volume are usually more predictable than the ones which determine value—and by "value" we often mean more than just "price." Given the long-term history of auto growth, for example, we could probably predict with a fair amount of precision that the annual growth in unit car sales would probably be in the order of 3 percent to 4 percent a year.

But if we were to look at the dollar sales history, we would be hard put to come up with anything nearly as accurate (see Figure 12-1). In addition to price projections, we would first need answers to (1) consumer preferences for big or little cars, (2) the amount of options and accessories these consumers would demand, (3) the impact of fuel shortages, (4) government tax policies, (5) mandated safety and pollution controls, and (6) new technological developments such as the rotary engine and battery-powered vehicles.

The volume part is reasonably predictable— depending on:	The value part is not — depending on such imponderables as:
Wealth of population	Inflation trends
Size of population	Consumer preferences
Age of population	Options and accessories
Family formation	Mandated safety and pollution
GNP growth	controls
Scrappage rates	Government tax policies
Saturation trends	Fuel shortages
	Technological changes

FIG. 12-1   Predicting auto sales.

Nobody could predict these latter factors with any degree of accuracy in the early 1970s, yet each and every one of them at that time promised to have a direct bearing on the dollar volume of auto sales over the following decade.

There's an added dividend to the deflation approach. In addition to zeroing in on physical volume, it forces you to take another good, hard look at the independent role played by price — a variable sometimes lost in the shuffle. Since the latter factor must be isolated to shift from value to volume, it presents a tailor-made opportunity to explore the important role played by price shifts.

All the above emphasis on physical volume and price by no means implies that the dollar figures should be ignored in statistical analysis. Far from it. There are occasions, for example, when it may well be the dollar rather than the volume data that are called for. Again, as in all statistical work, the problem at hand determines the choice of variables.

Harking back to the use of dollar sales, such figures, while clearly unsuitable for capacity planning, can become an integral part of any overall master plan involving future profits and cash flows. It is also this dollar figure which must be used when comparing a company's sales performance with that of individual competitors or the overall industry.

ADJUSTING FOR QUALITY

Just as price can distort dollar figures, so can quality changes. The point is that, if quality shifts, we are no longer measuring the same product—so some sort of adjustment must be made. Thus, as cars are made safer, more reliable, and less likely to pollute our environment, they are not the same breed of animal they were a few years ago. As such, we must factor out any price changes due to these variables, because, by definition, price indexes must measure changes in the prices of items of constant quality.

And it makes good common sense. If you're getting a better car for your money, you would normally expect to pay more for it. This might become a bit clearer if you think of your quality-improved car as a bigger car. Just as you wouldn't compare the price of a compact in one year with the price of a luxury car in the next year, you would not want to compare the price of a car before quality improvement (analogous to the compact) with the price of the same car after quality adjustments (analogous to the luxury car).

Viewed from still another perspective—if you can now buy one machine instead of both the previous two because the new machine performs twice the work per hour, it is probably worth twice as much to you. Following through on this illustration, if the price remains unchanged, you are, in a sense, getting twice as much value or utility for your money. If previously you had to lay out $2,000 for two machines and now you have to buy only one at $1,000, you are, in essence, buying your needs at a 50 percent savings. In this case, statisticians might well decide to drop the price index for this particular category of machine by 50 percent.

On the other hand, if the manufacturer decided to double the price of each machine to $2,000, you as a buyer would be no worse off, because you would still have to shell out only $2,000

—the same as you did for two machines the previous year. In this instance price men would probably say there has been no price increase after adjusting for the quality (productivity) improvement. Thus the quality-adjusted price index would remain unchanged.

To sum up, when we deal with price indexes, we must look at quality as well as the nominal price. If the increase in both is approximately the same, there has been no "real" or "true" price increase as far as satisfaction to the individual consumer is concerned.

One of the big problems faced by index makers every year is to weigh the quality improvements in cars against the increase in suggested list price. And sometimes curious results ensue— with the list tag going up but the reported price index going down (because the value of the quality improvements outweighed the increase in list prices).

Many times the problem is complicated by the fact that quality deterioration takes place at the very same time that quality improvements are being made. Then the two conflicting trends must be weighed against the change in list prices to see if the "real" or quality-adjusted index is to go up or down.

This, too, often occurs in the auto market. In one recent year (1969), for example, improved safety equipment, engines, and ventilating systems were judged to be worth about $24 to the consumer. But at the same time a downgrading of warranties (judged to be a quality deterioration) was estimated as a $23 net loss to the average buyer. Result: compensating quality improvement and deterioration, so that the reported list increase (about $70 that year) was also judged to be the real or quality-adjusted advance.

Make no mistake about it, the difference between reported and quality-adjusted auto prices can add up to quite a sum in a few short years. One study, for example, found that after adjusting for quality improvements, prices of a competitive group of cars rose only fractionally from 1954 to 1960, although their unadjusted list prices rose by 34 percent.

The BLS has paid more attention to this factor in recent years and has received cooperation from automobile manufacturers in evaluating quality changes. While this may have reduced the upward bias in the automobile component, there are probably other commodities for which a similar but as yet unquantified bias still exists.

In any event, few statisticians will generally argue with the concept of filtering out quality change. Problems arise, however, in determining how quality should be defined and what constitutes constant quality.

The general consensus, however, is to follow the physical approach in defining quality changes. Specifically, changes in physical characteristics that affect safety, performance, durability, and/or comfort and convenience are classed as quality changes. Adjustments are made for these changes so that the index will be based on prices for the same or equivalent quality.

Finally, there's the problem of putting a dollar figure on these quality changes. The government, in working with autos, uses three types of approaches based on three different costs or prices.

Where possible they opt for producer costs—the auto maker's extra outlays for labor and materials. When these are not available, government statisticians turn to replacement cost prices, deflated for additional costs such as storing, wrapping, and shipping and the extra margin usually applied to replacement parts.

It should be added that quality adjustments are also deemed to have taken place when there is a change in optional equipment. Thus if some piece of equipment which was formerly included in the sales price is made optional, then a quality adjustment must be made to raise the "real" or "true" price, because now the consumer is getting less for his money. Similarly, if something originally considered optional is made standard, then a downward adjustment in price is considered appropriate if lists remain unchanged. Usually what happens in the auto industry is that an optional item is made standard—with the list pushed upward to take account of the increased cost. Result: lists go up but the quality-adjusted price index remains unchanged.

It was a similar line of reasoning that persuaded price-index calculators a few years back to raise the quality-adjusted price when warranties were cut. Government statisticians regarded this an an elimination of a previously "standard" service item. Ergo, a quality deterioration.

Not everybody, of course, is in agreement with the value placed on these quality adjustments. Observes one top union official, "The government gives the auto companies entirely too much for so-called quality improvement." Uncle Sam's statisticians, he adds, are "completely dependent upon the voluntary cooperation" of the auto concerns and thus "can't afford to antagonize the industry."

Consumer advocate Ralph Nader agrees. "The government is only equipped to handle quantitative data," he says. "When it comes to quality, they get it voluntarily or they don't get it at all."

At a Senate hearing, Mr. Nader once accused the auto companies of not telling price economists "what the deteriorations in any given year have been, such as in cutting down tire size or skimping on brake linings." He further charged that the government didn't have engineers equipped to evaluate the industry's quality claims, and that the auto companies usually prevented their engineers from talking to the government.

A GM executive retorted, saying that the government sometimes refused to make allowances for claimed quality improvements. For instance, he recalled that the government wouldn't count as an improvement larger front windows even though they "improved visibility."

Strange as it may seem, there is often considerable product deterioration as well as improvement. A spokesman for Consumers Union, for example, recently noted that there are several reasons for such quality deterioration. First, he cited a recent tendency toward reduced quality control at many factories. He blamed this in part on the annual model changes for appliances and other big-ticket consumer items. His conclusion: The pressure to get the new model out in time often makes it impossible for the producer to do very much in the way of quality control.

Tough price competition was cited by this expert as still another reason for poor quality. Producers, he said, are often under severe pressure to cut back on quality to reduce price. He cited black-and-white television sets, which recently came down in price but in which the absence of horizontal control knobs and knobs to adjust brightness levels was conspicuous.

Obviously, then, there is a dispute over the effect of quality changes in existing commodities and services. However, the consensus is that an upward bias exists in this area, though it probably does not exceed 0.5 percent to 1.5 percent per year.

Odd as it may seem, despite the quality distortion, there are many who would avoid making any index adjustment. These people argue that if you have to pay, say, $200 more for a car to cover improved safety and antipollution safeguards, this represents a $200 out-of-pocket increase for the consumer—and a $200 increase is a $200 increase no matter how you choose to slice it.

Following up on this, these critics of quality adjustment say that such corrections tend to confuse the buyers. If a customer is asked to pay $200 more per car while the government and the industry are loudly proclaiming no price increase, he begins to question the authenticity of indexes—for, to him, price and out-of-pocket costs are synonymous.

But this is hardly a cogent argument against the quality adjustment. The buyer must eventually realize that you get nothing for nothing. Clean air, safety, and reliability are things that he has been demanding for years. Now that he's getting them, he has to pay, but he should not confuse such payment with the basic car price. In short, John Q. Public must be made to realize that he's getting a better car. To compare the price of this better car with an inferior one of a few years ago is just another example of comparing apples with oranges.

The task of educating the consumer is not quite as difficult as it may sound. Thus we all know we are paying more for medical costs. But no one except the most naive would suggest he's getting less for his money than a few decades ago, when sickness and death rates were so much higher. The same goes for drugs. They may cost more now, but few would dispute that the money saved in terms of reduced illness and quicker recovery more than offsets the higher initial drug outlay.

SEASONALITY

You'd expect to pay more for strawberries in the winter than in the summer. And your department store counterpart would be in pretty serious trouble if he didn't sell more in the pre-Christmas season than he did in the slow summer months.

There's nothing mysterious about all this. These are the economic facts of life. But more times than not, this seasonal effect is forgotten when numbers are presented. The raw data on prices, sales, inventories, etc., are collected—and then wrongly compared with other times of the year when entirely different seasonal influences may be operating.

Again some kind of adjustment is called for—one that can isolate the seasonal swing and give the reader or listener a more accurate view of what is happening at any given time. Actually, the objectives of seasonal analysis are two in number: (1) to find out what the seasonal swing actually is and (2) to gauge the true underlying movement of the figures being presented — the trend uncluttered with seasonal distortions.

Take the first goal — isolation of the seasonal pattern or index. It's a key bit of information, for by knowing at what time of year things are slow or booming — or cheap or expensive — you can better maximize your buying strategy — and save considerable money, time, and effort in the process.

This is certainly true in the case of auto prices. Thus the month-to-month seasonal index shows some surprisingly sharp swings. Uncle Sam's statisticians figure this amount at a bit better than 5 percent over the year (see Figure 12-2) — with the

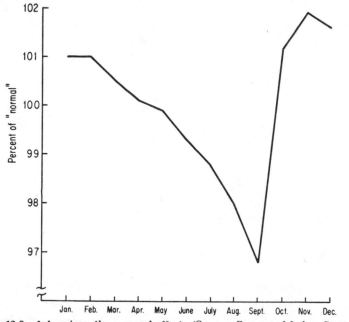

FIG. 12-2   **Auto prices, the seasonal effect.** *(Source: Bureau of Labor Statistics.)*

low point in September (just before new-model introduction time) and the high point in November, when new models are beginning to roll off the production line.

And even this may be on the conservative side. For price indexes don't always capture all the discounts during periods of weak demand. As such, the swing from seasonal high point to seasonal low point could well be closer to 10 percent.

But a word of caution. Buying at the September low point need not always be in your best interests. That's because the car will be a year old (model-wise) within a month or two—and hence, if it's traded in within two or three years, the savings on the purchase price may be more than offset by a lower trade-in price. People who keep cars for five years or more, of course, are scarcely affected by this trade-in consideration. For them a September purchase would always seem to be in order.

The second major objective of seasonal analysis—the removal of the seasonal pattern from the data to determine the underlying trend—also deserves some further comment. Statisticians and economists usually refer to this as *seasonally adjusting* the data.

This can be accomplished rather easily—witness the spate of government and private reports that now come out regularly on a seasonally adjusted basis. Nevertheless, there is a tendency on the part of some people to ignore such seasonal adjustment. Instead, they prefer to compare the data of any given month with those of the same month a year ago, thus canceling out any seasonal influence that might occur. But there is a serious weakness to this approach: it ignores the 11 months in between.

As one storekeeper put it, "Any approach must start with knowing how the recent level and trend of sales compare with those of the immediate past as well as with those of a year ago."

A department store manager tells this story of how his company erred seriously by using the year-to-year approach. The firm noted that its sales were running 5 percent ahead of the previous year's level. Comparisons made several months earlier showed that year-to-year gain was in the order of 10 percent.

Company executives, after looking over the available data, were not too happy but still felt that sales were in an uptrend. They drew up schedules calling for a continuing increase in inventory compared with levels of the year before. Much to their chagrin, they found that within three months they were actually running below the previous year—and they were stuck with a lot of costly inventory.

Hindsight revealed that if the data had been adjusted to take account of the seasonal pattern, the company would have noticed a change in trend (downward) before they gave the full-speed-ahead order. Use of seasonally adjusted data (data with the seasonal effect removed) could have saved a lot of red faces as well as a considerable sum of money.

A few words are also in order on how the seasonal adjustment is actually accomplished. Take the example of a retail store again. This time suppose you sell $100,000 worth of merchandise in July and $125,000 worth during the subsequent December. Question: Just how well are you doing? Obviously, some seasonal improvement would be expected in December, so your $25,000, or 25 percent, sales gain isn't quite as good as it looks.

The first step is to go to government sales statistics and find the seasonal index that best fits your sales mix. Assume the indexes for the two months in question are 90 and 110. In other words, you would expect July to be 10 percent under "normal" and December to be 10 percent above "normal."

If it isn't feasible to use one of the government-supplied indices to seasonally adjust, you may want to construct your own seasonal index. This is a bit more involved, but it can be done relatively simply and cheaply with the use of a computer.

The same approach would, of course, apply to the seasonal

adjustment of prices. Again the government has come to the rescue, with literally hundreds of seasonal index factors similar to the one on autos discussed above. These have proved indispensable for removing the seasonal price factor in any month of the year. The formula to apply is:

$$\frac{\text{Market price}}{\text{Seasonal factor}} = \text{adjusted price}$$

Use of adjusted rather than market prices will eliminate price movements stemming from normal seasonal demand, production cycles, weather, etc., and will spotlight changes due to new market conditions and sharp shifts in demand or supply. The hypothetical example below shows how this method works on the cost of fuel.

The price of fuel oil is $5 per barrel in July and $5.25 the following January. But we also know that normal seasonal influences pull up fuel prices in January. Specifically, previously calculated government fuel oil seasonal indexes for July and January are found to be 99.6 and 100.7, respectively.

QUESTION: How much of January's increase is due to basic underlying trend, and how much is due to normal seasonal influence?

SOLUTION: To sort out the influences, divide the reported market price by the corresponding monthly seasonal index factors:

$$\text{July} = \frac{\$5}{99.6} = \$5.02$$

$$\text{January} = \frac{\$5.25}{100.7} = \$5.21$$

The adjusted rise is 19 cents per barrel (or 3.8 percent). Thus about one-quarter of the reported 5 percent jump was seasonal, with the remaining portion due to underlying market factors.

One final note of caution. Take a close look at basic changes in the market before blindly applying last year's seasonal corrections. Automobile prices, for example, fluctuate around new-model-introduction time. So any change in the model-introduction date will have a significant effect on the seasonal variation pattern that year.

THE DOUBLE ADJUSTMENT

If we can seasonally adjust our data and deflate it for price changes, why can't we do both at the same time and come up with a seasonally adjusted, deflated series? Actually there's no reason why we can't. And so much of the most meaningful statistical information available now comes with this "double adjustment."

The procedure is essentially the same as when either seasonally adjusting or deflating. While we previously divided data by (1) seasonal adjustment factors, to come up with a seasonally adjusted series, and (2) price indexes, to come up with a deflated series, we now perform a double division—dividing the reported dollar amounts by both correction factors.

The big plus of this double adjustment is that it removes from the original dollar data two factors which can play hob with long-range planning: seasonal swings and price distortions. Left is a "clean" series—one which gives the reader or listener a clear-cut view of the basic underlying physical trend over time. To a large extent that's exactly what's needed to plot the future course of economic growth. No wonder then that one of the most widely watched economic statistics is seasonally adjusted GNP in constant dollars. It's far more meaningful than the actual dollar GNP, the seasonally adjusted GNP, or even the deflated GNP.

But the mere fact that GNP exists in so many versions gives the statistical charlatan the ammunition he needs to confuse and distort. As such, your best bet is to get the data in all four different versions (reported, seasonally adjusted, deflated, and seasonally adjusted-deflated).

This use of multiple versions of a given series is equally important in the private sphere. The businessman has to know (1) the seasonal effect, to plan near-term production schedules, (2) the inflation effect, to project cash flows and formulate pricing strategies, and (3) the combined effect to plan future facilities and permit month-to-month comparisons with such bellwether series as national consumption and output totals, which are usually expressed in terms of units, pounds, tons, etc.

Some analysts prefer to plot one series against the other to highlight the seasonal or price effect. Thus a seasonally adjusted-deflated series stacked up against a simple seasonally adjusted series will spotlight the price effect. Similarly, plotting the first series against a simple deflated series would project the role played by seasonal swings.

ANNUALIZATION

There's still one more possible, and often used, statistical adjustment. It is known as *annualization.* This is nothing more than the blowing up of weekly, monthly, or quarterly data into yearly totals. Thus if your supermarket bill rose 2 percent over a three-month period, we would say that supermarket prices were rising at an 8 percent annual rate. Translation: If prices continue to rise at the current rate, they will be up 8 percent in a year's time. Sales, inventory, and new-order figures also are often annualized.

The reason isn't too hard to find. This "inflating" procedure permits you to compare what has happened during a relatively short span of time with the movements of previous years. Then, too, we tend to think in terms of 12-month periods — and by reducing everything to this common denominator, we reduce the chance of confusion and standardize economic and business analysis.

But annualization can sometimes be misleading. Take the experience after the price freeze of late 1971. Under the more liberal controls then in effect, many sellers were able to boost prices—making for a bulge in both the consumer and the wholesale price indexes. But annualization of the rates of increase at that time was hardly justified. That's because many of the early post-freeze boosts were merely "catch-ups" of previously postponed increases. Then, too, some of the moves were one-shot affairs—with sellers using up all of their limited 1972 price-boosting authority in one fell swoop.

The critics who declared that the price stabilization machinery had gone haywire because of the then-current price performance were somewhat like a scorekeeper in a golf tournament who declares that a professional golfer is out of the running because he goes 1 over par on the first hole. The scorekeeper could say, "Look, at this rate, he'll be 18 over par and he might not break 90." He overlooks the probability that a pro golfer is not likely to maintain that average—and that, in fact, a good golfer is likely to pick up a birdie or two along the way to make up for his one mistake.

THIRTEEN

VARIATIONS ON A THEME

There's more than one way to skin a cat. Indeed, there are many ways.

And therein lies the problem wherever numbers and figures are quoted. No matter how many distortions and deceptions are exposed, the statistical charlatan is there—ready, willing, and able to come up with new ones. Anticipating them all is, of course, impossible. On the other hand, there are certain types or "families" of deceptions which crop up with disturbing regularity. Some, such as those involving phony statistics, improper use of averages, percentages, etc., have been explored above.

It is the aim of this chapter to put as many of the other types as possible under the spotlight—in the hopes that they will be recognized for what they really are: attempts to give the reader or listener a distorted view of the real world.

But in all fairness it should be pointed out that not all such errors are premeditated. Thus a distortion may possibly occur because the writer or talker is so immersed in what he's trying to prove that he "sees" only those numbers and figures that fit his purpose. Thus the liberal, hell-bent on welfare expansion, will quote reams of statistics to show how much good his proposals will do. But, in the same vein, he dismisses with a shrug the "negligible" cheating that such proposals might encourage on the part of welfare recipients.

Meantime, on the other side of the political fence, the conservative is so bugged by figures on how many dollars might be lost through cheating that he fails to see the economic and sociological pluses that might result from welfare reform.

Both men may be equally sincere. But because of their orientation and political persuasions, they are often unable to appreciate the valid statistical points of the other side.

But all the above may be too charitable. In the majority of cases the protagonist is probably well aware of both sides of the statistical picture, yet he chooses to ignore the side that doesn't fit — or even with premeditation distorts the numbers on his side to convince the fence-sitters.

A congressional debate a few years ago highlights this latter tack. The subject involved voting by black Americans. A liberal congressman from the Northeast took the floor and pointed to a county in Arkansas where 78 percent of the whites were registered but not a single black.

Shortly thereafter, the distinguished senator from that Southern state replied that the congressman was entirely correct in his statement, but had failed to note that only two blacks lived in the county — and that there was some question in the collective mind of the Census Bureau as to whether either of them was a black.

Coming closer to the consumer pocketbook, a car-leasing firm is quick to point out how, by leasing, you can save x amount of dollars by conserving on capital and y amount of dollars by conserving on maintenance costs—and generally ride around in a new-model car. What the firm conveniently forgets to point out is that, over the longer pull, the total outlay for the car will be larger under a leasing arrangement than under an outright purchase.

Again this is not to imply that the leasing approach is bad. There are pluses—advantages that some people are willing to pay for. But in all fairness, both sides of the statistical ledger—debits as well as credits—should be presented to the potential car user.

At this time, rather than enumerate hundreds of similar-type distortions, it might again be best to go into some of the broad-gauge categories of statistical legerdemain that always seem to be cropping up in one form or another.

APPLES AND ORANGES

Nearly everybody at one time or another has fallen into the trap of trying to compare the incomparable: a retail price stacked up against a wholesale price; a list price against a market price; wage rates against wage costs; the income of one nation against the income of another. The list could go on and on, but there's a common thread running through all of the above: in no case is there a one-to-one relationship involved. The magnitudes being quoted on each side of the comparison are essentially of different nature. To compare the apple on one side with the orange on the other is an open invitation to distortion and possible error.

Take the first example noted above—the making of comparisons on different levels of trade. It just can't be done—yet we all do it at one time or another.

Thus if food prices fall 10 percent at wholesale, we are likely to infer a similar deduction on the consumer level within a few months. To be sure, a large fall on the primary level will eventually result in a decline at retail—and that's why wholesale prices are used as a gauge to price changes on the supermarket level. But there's little a priori evidence to suggest that the change will be anywhere near proportional.

Indeed, the evidence is quite to the contrary. Much of any decrease that might occur at wholesale for example, would be siphoned off into advertising, packaging, marketing, and the like. And should wholesale prices rise—as they often do—the increase is actually magnified at retail because of these same cost pressures.

The statistics would certainly tend to back up this contention. In the early seventies a government study found that two-thirds of our food bill was earmarked for middlemen—the people who process, package, and sell the goods. In short, the farm price now plays a considerably smaller role than previously in the determination of the final supermarket level—in sharp contrast to a few decades ago, when the farm price accounted for over half of every grocery dollar.

All the above is in no way aimed at laying the blame on some "greedy" middlemen. The public obviously wanted the conveniences that have brought about the current price situation. In any case, it would be the height of folly to expect to pay the same per-pound price for frozen french fries as for potatoes in a brown paper sack.

The point to make here is only that there has been a shift in the farmer-supermarket relationship—one that is increasingly clouding the one-to-one relationship between prices on both these levels.

Moreover, level-of-trade comparisons are suspect on more than just the price level. Differing retail, wholesale, and manufacturing patterns are evident in most other business data as well. Profits at retail, for example, very rarely follow the producer trend—and certainly show a different amplitude of change. Nor could margins be compared on these two levels—if for no other reason than that the levels of investment and risk are quite different on each stratum. Much the same caveats can be given for comparisons of inventories, sales, and a host of other commonly used business yardsticks.

Another example of confusion over different levels of trade: A few years back (in 1971) the government announced that $3 billion worth of heroin had been seized around the world as the result of an American-inspired crackdown.

The truth: Like so many narcotics captures, the figure applied to the street price—the price after the drug had been wholesaled, retailed, diluted, cut, and packed in little bags and distributed down a line of pushers and receivers of payoffs. Under these level-of-trade transfers, a $10,000 haul at the top level can be ballooned into something like a $100,000 street price.

Why is the $100,000 price used? In many cases only because the bigger figure is bigger news. It also makes the police look good—when in fact all the transfers that might have ballooned the $10,000 figure into a $100,000 figure have never taken place.

Such blowups can't help but have their bad effects. For one, it might make the reader think the dope trade in his town had been broken simply because one seedy syndicate member had been arrested with $100,000 (street price) worth of heroin. The truth, of course, is that a gangster had been nabbed with a much smaller $10,000 investment.

There's another, even more insidious effect stemming from the use of the blown-up street price. It turns cheap hoodlums into imaginary millionaires, thereby giving the misleading impression that crime pays—and that this particular type of vicious crime pays exceedingly well.

Another type of apple-and-oranges pitfall: the irrelevant comparison. During the 1972 presidential campaign, a politically oriented editorial stated that while wages were rising at a 5½ percent rate, food prices had soared 25 percent. Specifically, the groceries that cost $20 when the previous administration took office in 1968 were now costing $25.

This is "apples and oranges" in its most blatant form; for the 5½ percent referred to the wage increase in one year (1972), while the 25 percent grocery figure referred to a cost increase over a four-year period (1968 to 1972). There is no reason why these two figures should have been compared—and it could leave the not-too-aware reader with the impression that the incumbent administration was bad for labor.

It may have been bad—and then again, it may not have. But certainly the numbers being quoted shed no light, either pro or con, on this important question.

Another glaring example of the irrelevant comparison was discussed in Chapter 10; namely, the confusion over wage rates and unit labor costs. As was pointed out at that time, the two are not synonymous, and any comparison of costs and prices must involve the unit labor cost concept. But hardly a day goes by that some corporate executive isn't bemoaning the fact that his firm's wage rates have gone up far faster than prices.

To compare these two magnitudes makes no sense—unless, of course, the purpose of the comparison is deception—for wage costs (which do offset prices) are a function not only of wages but also of output per man-hour (productivity). If a valid comparison is to be made, it must be one that pits prices against the combination of wage rates and productivity—that is, unit labor costs.

Then there's the typical market-versus-list confusion. This problem was touched upon in Chapter 11 on index numbers—because, while such yardsticks purport to measure true prices, government compilers in many cases are simply reporting the fictitious lists supplied to them by manufacturers, wholesalers, and retailers. The point to make here, however, is somewhat different: confusion between list and market price can often result in some costly mistakes—particularly on the consumer level.

Autos perhaps provide the most glaring example of the kind of mistake that can occur. It's no secret, for example, that there is a manufacturer's suggested list price for every model. But to assume that this bears any resemblance to the true transaction price is, of course, ridiculous.

Anyone who has ever purchased a car can attest to this. But for the few doubters, a recent Federal Trade Commission study should be the convincer. The agency checked the records of over 6,500 brand-new 1969-model cars—and, as might be expected, found that less than 2 percent of all American-made cars moved at list. To be sure, the percentage of foreign-made cars going at list was somewhat higher—but even here it still comprised only a relatively small portion of all foreign cars sold.

In short, the naïve buyer who looks at the showroom tag and then purchases a new car at that price is clearly being "taken." The fact that manufacturers preface the word "list" with the word "suggested" would also seem to provide further evidence that such prices are just the starting point for serious bargaining.

Nor is the usual difference between the list and the transaction price nominal. Again harking back to the FTC study, results indicated that the majority of the surveyed cars were eventually sold for anywhere from 10 percent to 20 percent under list.

That's a lot of money when an average purchase of upwards of $3,000 is involved. Just how much was also pointed up in the FTC survey. Almost 90 percent of the buyers managed to knock at least $200 off the original asking price. And in cases where buyers held the upper hand and where the purchase involved more expensive models, the "savings" ran as high as $1,000 or more. Equally significant, the average discount off list reported at that time came to something in the $500 to $600 range—something that could hardly be classified as chicken feed, even at today's inflationary price levels.

So far the discussion has been limited to cars. But similar list-transaction differentials exist in almost all appliance lines. Indeed, the success of discount houses in the 1950s was due in large part to their bringing out into the open—and offering to everybody—deals which had previously been made available, sub rosa, to a privileged few.

Another "apples and oranges" caveat: In today's modern world with its mixing of cultures and complex political alignments, there is an understandable desire to compare one's own standard of living with that of other nations. Figures are quoted almost daily in the press, for example, purporting to show how much we are ahead of other countries—and how the gap is either closing or widening.

These statistics do have their place—and they do provide rough yardsticks. But they're not nearly as accurate as some of the quoters would have us believe. Certainly any attempt to compare the GNP or wealth of one country with that of another is fraught with danger—for a variety of reasons, first and foremost of which is the use of international exchange rates.

There's scarcely an economist who today would admit that these rates are realistic for converting the purchasing power of one country into the purchasing power of another. In many cases, they're kept artificially high or artificially low for political reasons. Then what do you do for a country that has dual exchange rates—one rate that applies to one set of transactions and another that applies to a completely different set of transactions? Finally, it must be realized that exchange rates are oriented toward the items that enter international trade. But these are scarcely the same mix as the goods that a typical consumer or wage earner would be purchasing.

But exchange rates are only one, albeit an important, part of the overall problem. The reported incomes in each of the countries to be compared may be equally questionable. Clearly a good portion of the income generated in a poor, underdeveloped country never gets reported—primarily because the food the people grow and the clothing they wear never go through the marketplace. (According to accepted GNP and income-counting standards, a product must go through the marketplace if it's to show up in the dollar totals.) It's hard to put a dollar sign on the underestimation involved. But in some of the more undeveloped areas of the world, the actual or true income and GNP generated may run upwards of twice the official or reported level.

There's still another possible error involved in making international living standard comparisons. It involves the assumption that the price levels prevailing in one nation will be relatively the same as in another. We may look at the income of, say, an undeveloped Asian or African nation and say that no one could possibly live at that low income level. What we're forgetting, of course, is that in such a country the price of basic staples, such as rice and corn, is very low—much lower than it would be in the United States. So it takes very little for these people to subsist on a minimal diet—much less than it does here. It is only when you begin to look at sophisticated goods that the very poor are shut out of the market.

Next time you're abroad, test this out for yourself. Go into the native store, and you'll be surprised to see how very little it takes to live on—provided, of course, you're willing to eschew the American way of life, with its cars, appliances, beefsteaks, and other amenities that are commonly taken for granted.

MIX MIXUPS

When talking in big numbers or totals, there is often a tendency to ignore the groups and subgroups that make up the aggregate. But anyone who is forecasting does so at his own risk—for seldom do the parts that make up the whole remain in fixed relationship to one another. Thus we might be selling 25 percent more autos now than, say, 10 years ago. But because of a shift toward smaller, more compact cars, it would indeed be an error to assume that profits or dollar sales (assuming no inflation) have gone up by a like amount. A rundown on some of the more common types of these composition pitfalls follows.

1. *Shifts in the price mix.* The auto example given above is only one of a series of such changes that are always occurring. Sometimes even the experts are shocked by such mix miscalculations. Take the example of a few years ago, when we thought imposition of quotas might be the answer to the then-existing flood of steel imports. During the first year of the pact (1969) it was stipulated that tonnage was to drop from the 18 million-ton level of the previous year to something around 14 million tons. This, in turn, was expected to result in a sharp improvement in our balance of trade.

But look at what actually happened. Overseas mills obeyed the letter but hardly the spirit of the law. They cut tonnage as decreed, but substituted much more expensive stainless and tool steels for the previously shipped carbon types. Result: Imports in dollar terms remained high—so high, in fact, that the quotas had virtually no impact in curbing our dollar outflow.

The above was, of course, a premeditated attempt to get around the ruling. But the value-volume mix pitfall is ever present; for every time the consumer upgrades or downgrades his product mix, he is, in effect, changing dollar totals. If nothing else, it has become extremely risky to assume that the product mix will be repeated over an extended period of years. This, incidentally, is why most economists and business analysts recommend that value and volume forecasts be done separately. It's the rare case when the two will move up or down in tandem— even in cases where the basic price levels remain unchanged.

2. *Comparing unlike mixes.* There's still another aspect to the mix problem: the tendency on the part of many to compare unlike mixes, and then to draw unwarranted and often misleading conclusions from the results. A case in point: Much has been made of the income disparity between blacks and whites—with

the conventional wisdom attributing most of this to prejudice and unequal opportunity. Nobody would deny that these factors have played a role. But a moment's consideration would reveal another possible reason for the gap: the geographical location of the blacks. A disproportionate number of them live in the rural South, where incomes are low for everybody — whites and blacks alike.

On the other hand, if you examine white and black incomes on a more proportional basis (say, include x percent rural white and the same x percent rural black), the conclusion is quite a bit different. Under this controlled experiment approach, close to half of the gap between whites and blacks disappears. Put another way, if the black population were distributed about the country as the white population is, the income gap wouldn't be nearly as impressive.

This kind of mix error isn't limited to the social sphere. The same mistake can appear in business as well. The union that compares income in two factories and finds that earnings in one are much lower than in the other may be guilty of the same distorted thinking. The first factory, because of the nature of its work, may have 20 percent supervisors and 80 percent blue-collar workers. The second may have a 10 percent to 90 percent breakdown.

It is clear, then, that the first factory will have a higher overall average income, even though the wage rates for the same type of work in both factories are the same. In such a case, to conclude that workers in the second factory are being discriminated against is obviously wrong.

The lesson to be drawn from both the above illustrations: Composition or mix plays a major role in influencing average levels.

Another example — this one also drawn from the social area. A few years back proponents of "halfway houses" (private institutions designed to help rehabilitate ex-convicts) attracted a lot of interest—because of the seemingly valid claim that this approach was proving far more successful than the use of prisons in rehabilitating convicts. Thus at that time some halfway houses claimed that fewer than 20 percent of their residents ultimately returned to crime (versus a reported 60 percent recidivism rate for all ex-convicts).

Even if you assume that the 60 percent rate is correct (something open to serious question), there's the fact that the rate of success of halfway houses was probably inflated by their rejection of bad-risk convicts. Many such houses, for example, screen out drug addicts, sex offenders, and other bad-rehabilitation risks.

In short, the convict mix of the typical halfway house is not the same as the overall convict mix. To assume it is, by comparing rates of recidivism, is to invite disappointment later on. As one prison official put it, "Perhaps a 5 to 7 percent reduction in recidivism is being accomplished by use of halfway houses"—something far below the claims made by the houses themselves.

3. *Changing mix and inflation.* A fair amount of our inflation over the past few years cannot be attributed to a sudden rise in costs, but rather to a shift in demand toward areas where costs have traditionally moved up at a faster rate. Specifically, more of our dollars now go to services, which have always tended to rise faster in price than products because it is harder to increase the productivity or efficiency of service-oriented industries. A barber can give only a certain number of haircuts an hour, a heart surgeon can perform only a certain number of major operations each day. We wouldn't want it otherwise. But the net result has been to give a greater weight to service industries and hence an added push to the price level.

One we recognize this, we realize that to keep the average price increase at levels of one and two decades ago would require a much more stringent anti-inflationary policy. Or perhaps we should accept a higher rate of inflation as the price of satisfying our voracious appetite for services.

4. *Fallacy of composition.* There is a "gut" feeling—completely unwarranted, as we soon shall see—that what is good for the individual is good for the country, and, conversely, that what is bad for the individual is bad for the country.

If you are one of the many who are hooked on this "fallacy of composition," consider the following: You read in the papers that a recession looms on the horizon, so you rightly cut down on discretionary outlays. But do you ever stop to think what would happen if everybody acted this way? For one thing, sales would fall off, inventories would climb, and eventually this would be followed by production cutbacks and layoffs. So here you have a case showing that if each individual acted in his own self-interest, it could well prove disastrous for everybody—with such "good" individual actions precipitating the recession that everyone had feared in the first place.

Happily, most people in government recognize the fallacy of this type of thinking. That's why when talk of a possible recession crops up, Washington officials immediately start to pooh-pooh it—for only by rebuilding confidence can the government prevent a nebulous fear from becoming a very real fact.

Budget balancing is also subject to this fallacy. It's argued by some that a government that runs deficits year after year is heading for bankruptcy. What these people forget is that a government is not a person. When a government goes into debt, it represents nothing more than a "book" transfer—the taxpayers (one group of taxpayers) owing bondholders (another group) money. It's analogous to a wife owing her husband money. As long as they remain a family unit, there's little danger of insolvency—as long as the husband can meet the bills.

AVERAGE COST IS NOT PRICE

Mix changes sometimes contribute to misleading price conclusions. Take the example of a few pages ago involving the shift from larger to smaller autos. Obviously this has tended to bring down the average car outlay—or at least has kept it from rising as much as it might have, say, over the past decade.

But this would not be a measure of the "true" price trend over the same period. The latter concept refers only to a change in price of a given mix of cars over a period of time. Clearly, the same type of car costs more today than a decade ago (even after correcting for upgraded quality). In short, the price curve has risen considerably faster than the average outlay curve over the last 10 years. And to look only at the latter could give a misleading impression of inflationary pressures in the auto industry.

But this is not to say the average outlay or unit cost figure is valueless or wrong. It is misleading only when used as a yardstick of inflation. On the other hand, if you're measuring the role of auto expenditures vis-à-vis consumer income, the average outlay or unit cost would seem to be the more appropriate yardstick.

The same dilemma appears where imports and exports are concerned. Up to the early 1970s, the government published only unit-cost trade indexes. But analysts glibly translated this into "price"—ignoring the fact that the mix of imports and exports is constantly changing. In such case there is no prior reason to suppose that the actual rise in average unit costs over recent years is synonymous with "price" change.

Given worldwide inflation, prices did probably go up. But looking at the unit cost index with its changing mix, it's hard to say whether the "true" price rise was the same as or more or less than that indicated by the unit cost trend.

OMISSION VERSUS COMMISSION

In many cases the figure finagler can get his point across without any outright tampering with the actual statistics. All he needs to do is conveniently forget to present all the facts. In some ways this is even more effective than any blatant juggling of the figures, for the miscreant can always plead that he didn't want to burden the reader or listener with too many dry statistics. But the impact is the same whether the deception is covert or overt. In either event it represents a premeditated attempt to leave the reader or listener with a distorted view of the real world.

One of the most common of these ploys involves the failure to relate a number to a proper yardstick. Thus the person who points out that more accidents occur during the day than during the night is telling us little—except leaving us with the erroneous impression that it might be safer to drive at night. In this case, the obvious yardstick for putting the accident figure into proper perspective—the number of people who drive during these two periods—has been left out. Similarly, the industry that claims it pays more corporate tax to Uncle Sam than another industry is following the same "omission" tack. For the pertinent question is not whether the tax is large or small, but rather what it is in relation to the income generated by the industry. It would be a sad day, indeed, if the auto industry paid less taxes than, say, the relatively small leather or brick industry.

Such omission-type distortions often appear in statements by industry trade associations. One recent example: The petroleum industry put out a release loudly proclaiming that it paid its average worker 50 percent more in fringes than was paid to the average industrial worker. This sounded impressive until it was pointed out that the average salary level in the petroleum industry was also 50 percent above the national average. When this additional, but key, bit of information came out, it was clear that, on a percent-of-wage basis, the petroleum industry was no more liberal than any of its fellow industries.

As might be expected, politicians are equally adept at omitting the appropriate yardstick. A conservative candidate will make much ado over the fact that our national debt is growing by leaps and bounds, that it has gone up $x billion over the four years since his opponent took office. He could or could not have a point—but one would know only if the necessary yardstick (GNP) was also available. That's because debt, like everything else in this world, is relative. A higher GNP (analogous to a higher income) would certainly permit the carrying of a higher debt. The point is, how much has the debt increased relative to this increase in GNP? Know this, and you know whether to be concerned or not.

This is quite obvious when viewed from an individual vantage point. A bank would have fewer qualms about lending a millionaire $1,000 than, say, a ghetto dweller even $100. Again it's not the debt, but the debt in relationship to the ability to pay it back that counts.

The communications industry is another perpetrator of the "omit-the-yardstick" ploy. A newspaper or TV commentator may make much ado about the fact that inventories went up $x billion. What he will forget to say is that business activity went up by a like amount so that the stock/sales ratio—the only meaningful measure of inventory level—remained unchanged.

The only difference between the people of the news media and other people who conveniently forget to include pertinent yardsticks is the basic reason for the deception. Most users of this ploy want to deceive. Our "friendly" news commentator only wants to startle, to make the statistics seem newsworthy. He cares little whether the change is up or down. His criterion: Make it big enough to make the reader or listener sit up and take notice.

Finally, let's not forget the government. The party in power may make much of the fact that prices rose only 5 percent during the current year—1 percentage point under the previous year. This sounds great—unless you take the time and effort to find out that this 5 percent rate is still above the 3 percent "norm" of the past decade.

At still other times, the omission involves a qualitative fact connected with the number rather than the number itself. A few years ago, for example, the GNP price level suddenly leveled out after more than a year of steady rise. But the reason was a one-shot change in the composition of GNP components—not any basic decline in the components that made up the average. Did the government publicize this fact? Hardly. The only mention was in a footnote at the bottom of the report (a place almost everyone was sure to ignore).

This error of omission came out only three months later when the next GNP price report showed a significant advance. Then the administration was quick to point out that the previous quarter's leveling off was a statistical quirk—and that therefore the price situation had not really deteriorated. That's "having your cake and eating it" with a vengeance.

People antagonistic to a certain approach or program also often fail to cite the pluses as well as the minuses. The controversy over auto emission–control devices comes to mind. In early 1973, opponents of such devices were quick to point out that this would mean about a 7 percent increase in fuel consumption—at a time when we were facing a growing gasoline shortage.

What these same people (the auto companies) conveniently forgot to mention was the fact that the luxury-oriented Detroit approach was wasting a lot more via (1) large cars (a 3,500-pound car uses 14 percent more gas on the average than a 3,000-pound car), (2) air conditioning, which eats up about 20 percent more fuel, and (3) automatic transmission, which on the average results in about a 6 percent loss in fuel efficiency. If the auto firms were serious about conserving fuel and people's money, they would dispense with these non-necessities (the savings in fuel consumption would come to nearly 40 percent far outweighing the 7 percent loss entailed by antipollution devices urgently needed to keep us all from choking to death).

THE FALLIBLE PROJECTION

Anyone attempting to forecast tomorrow's weather, next month's income, or next year's auto sales is, in a sense, sticking his neck out; for the chances of going off the track are many and varied. If nothing else, the assumptions made at the time of the projection are always subject to change (see Chapter 8). But forecast errors can stem from a variety of other causes, too, including:

1. *The changing role of determining variables.* What's not significant today may be significant tomorrow. And vice versa. There's nothing really profound in the above statement. But it's a fact of life that can play hob with forecasts that go well into the future. Thus if prices that have been rising are falling within a narrow range, they may not be an important demand-determining variable. But let a major shift take place, and price becomes crucial.

Metals have found this out the hard way. In copper, a sudden doubling of the price in the 1960s found users deserting the red metal in droves in favor of aluminum and plastic substitutes. And in the early 1970s an almost-as-impressive run-up in zinc tags found more and more die casters opting for plastics.

Another example—this one drawn from the farm sector. Temperature may not have been a major determinant of crop yield during a long period of time when rainfall was normal. But let precipitation suddenly drop off one year, and temperature immediately becomes a crucial factor.

Sometimes a small technological change can make a world of difference, changing a time-tested forecasting equation almost overnight. Thus the rapid development of miniature electronic circuitry in the early 1970s suddenly opened up a brand-new market for small pocket-sized calculators. By 1973 it seemed as if every single junior executive was clamoring for his own toy machine—in most cases, more for prestige than for any sophisticated arithmetical calculation that he might be called upon to perform. Result: Calculator sales zoomed far above what might have been expected on the basis of profits and income.

2. *Short base period.* Since forecasting is based primarily on historical relationships, the past record must be of sufficient

length to assure validity. One expert, while he admits that no a priori or empirical answers to the questions are available, points out that it would be "statistically unreal or unreasonable to use one or two time units such as years, months, or days for the purpose of projecting five or more equivalent time units." This expert goes on to say that "a rough rule of thumb employed by some analysts is that the length of the base period should be at least equal to the term of the projection period." Only then can the results meet the rigid mathematical criteria for a forecast that is both valid and useful.

3. *Extrapolation beyond the range of experience.* This can best be described by a simple hypothetical example. Assume that historical data reveal an excellent relationship between small-appliance sales and family income. Specifically, sales tend to rise 1 percent, say, for each $100 rise in annual income. Furthermore, the relationship has been derived from past experience, in which income has varied from $5,000 per household to $7,000 per household.

So far, so good. But what happens if income spurts to $8,000? Does the same relationship hold? One can't be absolutely sure. It is a little risky to assume that the same 1 percent rise per $100 of income will prevail when income goes much above the $7,000 mark.

4. *Trend perception.* The temptation to project short-run developments into long-run trends can often lead to error, or at least to a difference of opinion. For example, some see the increased purchase of services (relative to goods) in recent years as a trend acceleration. Others, pointing to a similar development in the 1920s, say NO. They feel that reviewing history over a sufficiently long period of time reveals no trend acceleration, but only two bumps on an otherwise smooth trend.

The same difficulty has evolved around productivity estimates. In the early 1960s manufacturing productivity was increasing at 4 percent or more per annum, compared with the 2.5 percent to 3 percent rate of the previous decade. Some analysts claimed that was a new trend. Others, pointing to previous spurts which later fizzled out, said NO. Who was right? It will probably take another decade to get the answer.

5. *Estimates on estimates.* This type of projection is generally associated with problems relating one variable to another. Thus, in predicting next year's sales, one might have to make an educated guess about next year's income. This means that the normal amount of error in projecting sales must then be multiplied by the possibility of error in projecting income (the sales-determining variable). With two areas of possible error, the chance of variation is substantially expanded.

As a rule of thumb, if it proves almost as difficult to estimate the determining variable as to estimate the variable to be predicted, then the correlating technique may not be the best way to tackle the problem.

There is also a compounding effect when the original figure and the final result have been estimated incorrectly. For example, a 10 percent overestimate in the determining variable, when combined with, say a normal 10 percent projection error could conceivably yield an error as large as 21 percent (110 percent x 110 percent).

FOURTEEN

FINANCIAL PITFALLS

Everybody fancies himself an amateur psychiatrist. Similarly, everybody thinks of himself as an astute financial analyst—able to ferret out the few really good stock opportunities from the run-of-the-mill and bad ones.

The only trouble is that, just as in the case of psychiatry, things are seldom what they appear to be. If there's any doubt on this score, think of all the surefire market tips you've received—backed up by supposedly solid facts and figures—only to see the stock in question dip sharply during the ensuing months. Firm A may show bigger sales, faster growth, and higher earnings than Firm B—and still turn out to be a far less desirable candidate for investment.

In any case, it behooves any investor to at least look at some of the crucial financial yardsticks that signal financial health or the lack thereof. These should certainly include measures of liquidity, cash flow, profit margins, and inventory turnover. A look at the price/earnings (P/E) ratio is also a must.

While detailed balance sheet and profit and loss statement analyses are beyond the purview of this book, a brief discussion of some of the above-mentioned yardsticks would seem to be in order, because of the distortions and misunderstandings that often arise. Other areas that suggest closer examination at this point include (1) outright attempts to change or distort the financial picture and (2) the basic inadequacies of financial statistics as an investment gauge. Examining each of the above-mentioned categories separately:

YARDSTICKS AREN'T EVERYTHING

All but the most naïve have probably discovered that the measures discussed above are not infallible. Few people have made their fortunes by playing the market strictly on ratios. A far greater number have probably "lost their shirts" by taking this tack.

Part of the problem can be traced to the unpredictable psychology of the marketplace. But equally important, these ratios are only outside manifestations of an extremely complex business enterprise. As such, they can't even come close to telling you all you need to know to make an intelligent business decision. If it were all a question of knowing ratios, then the entire security industry would be superfluous. We all could be our own security analysts.

Finally, the actual interpretation of these ratios can also sometimes be faulted. Great care, for example, must be exercised to avoid the old "apples and oranges" pitfall. As pointed out in an earlier chapter, profit margins of Firm A are valid only when compared with Firm A's own industry—for industry margins tend to differ significantly. The same might be said about liquidity ratios and inventory turnover rates.

Consider for a moment the latter measure. True, a turnover ratio (the cost of goods sold divided by inventory) supposedly shows how successful a firm has been in controlling inventories. And certainly a high turnover rate is preferable to a low one. But note these eye-opening differences: Chemical and petroleum companies traditionally maintain turnover rates in the seven- to eight-times-a-year range. On the other hand, machinery firms are lucky to turn over stocks three or four times a year. But there are good solid reasons for these differences. So again, interindustry comparisons are virtually meaningless.

Even cash flow measures must be examined carefully. Thus if a firm shifts from normal to accelerated depreciation, cash flow (defined as net profit plus depreciation) will tend to take a sudden spurt. Net profit, the other component of cash flow, can change for a variety of noneconomic reasons, too.

Price/earnings ratios (the market value of a stock divided by a firm's net earnings) are equally vulnerable. To be sure, normally, the lower the reported P/E ratio, the more attractive the security. But, like any rigid statistical yardstick, the P/E ratio must be applied with caution.

Note, for example, that a P/E ratio is figured on the profits reported over the previous 12 months rather than for the latest week or month. Such being the case it's important to examine all events over that time span to see if something unusual hasn't influenced earnings. Specifically, extraordinary events, such as a strike or buying hedge, can distort results—the former raising the P/E ratio above normal levels and the latter reducing it below normal. In neither case would the reported P/E ratio be telling the true story.

But even in the absence of such distortion, caution is advisable —and certainly the P/E ratio should not be the only factor in a decision on whether to buy or sell a certain security. Clearly, any blind selling of growth stocks when the P/E ratio reaches even as high as 30:1 can be disastrous. The history of such growth companies as IBM and Xerox proves this point beyond a shadow of a doubt.

Another P/E caveat: What passes for a market-acceptable ratio in one period may be completely unacceptable in another. Reason: Stock market fashions may change—and an industry that today attracts a high price because of its glamour may not do the same tomorrow. Thus in the early 1960s, boating, bowling, computers, and conglomerates were all glamour issues with high P/E ratios. But by 1970 all had lost favor with investors, with some P/E ratios in these areas dropping from over 40:1 and 50:1 all the way down to 10:1—even in cases where the companies involved lived up to their growth expectations.

But let's not throw out the baby with the wash water. There are some general P/E guidelines that can be useful to the investor. Other things being equal, for example, we would expect a growth company with consistently high profits to carry a higher P/E than a lackluster, stagnating outfit. And with good reason. The former company has better potential and the investor is generally willing to pay a premium in the hope that this potential will materialize.

To sum up, P/E and other financial ratios are useful and do perform a function. Certainly, other things being equal, a favorable ratio is a lot better than an unfavorable ratio. The only problem is that other things are seldom equal. Thus, even when free of deception, financial ratios can sometimes lead to the wrong decisions and choices—unless, of course, the investor is willing to take the necessary time and effort to look for the actual forces behind the numbers being reported.

OBSCURING THE TRUTH

In this section, the spotlight will be placed more on outright deception and/or premeditated moves to leave the reader with a misleading impression, and less on simplistic misinterpretations of data on the part of the reader or investor. Such deceits have not been all that uncommon. Indeed, up until a few years ago many financial reports were little more than fiction. It has been only in recent years that some degree of honesty has been injected into balance sheets and profit and loss statements.

Part of this start toward fuller and more honest disclosure can be traced to government prodding. But equally important has been the realization on the part of many firms that much of the accounting sleight of hand and secrecy just wasn't paying off any longer.

But there's still a long way to go before anything even remotely resembling full honesty can be expected. This is so because old myths die hard. Thus one of the reasons why some old-line firms are reluctant to reform is the hope that they can put one over on the tax collector. But in most cases this is little more than wishful thinking. The Internal Revenue Service knows all the tricks by now, and the more blatant forms of skullduggery are quickly picked up.

Other firms have thrown in the sponge as far as Uncle Sam is concerned. Their tax returns are usually above reproach. Instead they concentrate all their fakery in their annual reports, which generally face much less auditing than their tax returns.

The question is why they continue their old habits here. Certainly the belief that they can hide the facts from prying union officials would seem to be untenable. Indeed, some union heads actually sit on the boards of some corporations today. And even when they don't, the large national unions usually have enough high-paid talent on their staff to be almost as astute as the Internal Revenue Service. Upshot: If corporations think they're fooling the unions by clouding up their financial reports, they're sadly mistaken. Chances are unions today know more about the financial conditions of the companies they're involved with than most shareholders—and that goes for fairly big-sized shareholders as well as the spate of small ones.

There are still other firms that continue to feel that full disclosure will put them at a disadvantage vis-à-vis their competitors. But there's precious little evidence to back up this belief either. Thus some of the fastest-growing firms have been most open in terms of financial reporting—and they've been located in highly competitive areas as well. When all is said and done, it is the best-managed, most innovative firm (not the most secretive one) that invariably comes out on top.

Fear of stockholders could be another reason behind corporate reluctance to reveal all. Some top executives, while publicly espousing full disclosure, privately admit to their doubts about giving shareholders the full picture. They fear, for example, that too rosy an earnings report (even if justified) could lead to stockholder pressure for higher dividends—thus making it more difficult to set aside adequate sums for reserves, research, and expansion. And on the other side of the coin, they feel that a bearish report (again even if justified) could spark a revolt against the current board of directors.

With all these diverse pressures, no wonder corporations often resort to multiple depreciation methods. This approach has gathered steam in recent years because of the need to show one face to Uncle Sam and another to investors and stockholders.

The approach here is quite predictable: a preference for (1) straight-line depreciation in public reports (a technique which reduces costs and hence raises profits) and (2) accelerated depreciation in tax reports (which raises costs and hence reduces profits and the tax burden).

No wonder, then, that in recent years there has been a shift from accelerated to straight-line depreciation in public reports. And in some cases this has resulted in significant differences in apparent profit performance.

Republic Steel, for example, in late 1968 noted that because of this change net income for 1968 was not strictly comparable with that for 1967. Specifically, the firm said that the new straight-line depreciation had yielded for the first nine months of 1968 a net earnings figure of $65 million, well above the $50 million of the comparable 1967 period. But Republic then quickly added that $9 million of this increase was due solely to the accounting change. In other words, only $6 million out of the $15 million profit gain was due to real factors. The rest was little more than accounting sleight of hand and had to be subtracted if any meaningful comparison with the previous year was to be made.

Individual company statistics also are often distorted by mergers or the purchase of an additional operating unit or two. The resulting addition to sales or profits makes comparisons with the previous year's performance extremely risky unless the analyst is aware of the changed corporate mix. The best way out of this dilemma, of course, is to compare performance of the same operating divisions over the two periods.

A similar situation crops up when a firm decides to sell off an operating unit. More often than not, this action is prompted by the unfavorable profit performance of the division involved, so that while an uncritical look at before-and-after sales and profit totals might suggest a decline, in actual fact that company may have strengthened its underlying financial position.

Other key questions one might explore when analyzing financial statements: (1) Has the corporation consolidated the sales and earnings reports of all its subsidiaries? (Obviously it should, to give you the entire picture.) (2) How does it value inventories? (The so-called LIFO and FIFO methods can make a world of difference on the bottom line.) (3) Have charges for advertising or research and development been deferred? (If so, when these charges are ultimately deducted, profits could plummet.) (4) Does the outfit have a lot of bonds due for redemption? (If so, new bonds at higher interest rates may have to be issued — something that could adversely affect future earnings.)

The first of the above questions deserves some further comments—for nonconsolidation is often premeditated. This approach, for example, can often hide losses by fictitious sales to subsidiaries. One spectacular case a few years ago showed that management had juggled the books to the extent that the firm, capitalized at only $20 million, had piled up debts of $320 million before it went broke.

Accumulation of hidden reserves is another popular distortion. This practice often enables management to keep unaccounted-for financial resources to meet later emergencies—such as making dividend payments in bad years. One way of hiding such reserves is by deliberately undervaluing assets. A firm may, for example, carry building and property on the balance sheet at a nominal $1 value. Then, too, some outfits hold substantial real estate or stock as investments, but one would never know from their balance sheets how much these assets have appreciated in value over the years.

Other motives can produce still other ways of doctoring up financial statements. For competitive reasons one firm may fail to reveal part ownership in another. Another popular way of concealing assets is to undervalue inventories. Still other firms in financial hot water may elect to put a high value on a substantial volume of unsalable goods—or even to treat nonrecurring income from sales of assets as if it were part of normal operating profit.

The list could go on and on. But the above should be sufficient to convince the reader that things are not always what they seem to be.

A FLAWED YARDSTICK

There are actually two aspects to the validity of financial statistics. The most widely publicized and obvious one concerns the type of deceptive practices outlined above. But equally important is the less-known fact that financial statistics, for all the lavish attention that has been heaped on them, are basically limited analytical tools. It is possible, for example, to have a perfectly honest and complete financial dossier on a corporation and yet be woefully uninformed as to whether the firm has been maximizing its potential.

Below are just a few of the popular financial concepts that have been taken for granted but which in many cases raise more questions than they answer:

1. *Net profit.* Strange as it may seem, two firms with the identical investment input, cost, and sales can end up with different net profit positions. The problem stems from our arbitrary taxing system—a system which has one set of tax rules when money is raised in the bond market and another set when money is raised in the equity (stock) market.

It all reverts to the fact that (1) interest paid out of revenue (to service the bond loan) is considered a cost of doing business while (2) dividends paid out to stockholders (to service the stock investment) is not considered such a cost. Thus a firm taking the bond route will find itself with higher "costs" and hence a lower net profit, even though everything else about the two firms is identical.

The difference can be substantial. Take the hypothetical case of two firms in the same line of business—one gets its capital via a $1 million bond flotation while the other gets its through a $1 million stock flotation. Both firms make the same product and incur the same manufacturing costs—and wind up with earnings (before interest) of $100,000 each.

But this is where the similarity ends. The firm with the bond issue which calls for 6 percent interest would have to pay bondholders $60,000—leaving a net book profit of only $40,000. But the stock-financed firm, even if it paid out $60,000 in dividends, would show a net profit of $100,000—because this dividend payment is not regarded as a cost. Upshot: Two equally well-run firms end up with completely different profit because of the vagaries of our tax system.

The alternative methods of depreciation discussed in the

previous section complicate the picture even more. Depending on any number of optional depreciation techniques, the profit of each of our two firms could be higher or lower. So still another arbitrary decision, sanctioned by our tax system, makes it yet more difficult to pinpoint "true" profits.

2. *Implicit costs.* The above examples can, of course, be traced back to the different ways of costing certain operations. But going beyond that, confusion can result from the fact that even if we agree that interest payments are a cost, we recognize only explicit payments, ignoring the sometimes equally big implicit payments.

Inventory carrying charges perhaps best illustrate the point to be made. If we borrow $10,000 at 6 percent to finance our inventories over a given year, we will show a cost figure of $600 on our books. This is obvious and can be defined as the *explicit interest cost.* But what if we had $10,000 excess cash in our tills? We might bypass the bank and use this $10,000 to finance our inventory needs. This time around, however, we would show no interest cost on our books because we used our own money. But we could just as well have put the cash reserve into a bank, where it would have earned $600. So, in a sense, by using our own money we were sacrificing $600 in potential interest income. This sacrifice cost the firm $600—but nowhere in our accounting rules do we see any provision for including this sacrifice or implicit cost when determining profit.

One can speculate on why only explicit costs are permissible deductions. Many feel, for example, that measuring such implicit costs would be difficult (Who's to say what your money would earn outside the firm?). But this hardly gives the alert manager any reason for ignoring this cost. As such, it is part of his job to consider the cost of money when determining optimum inventory levels—whether such money is generated inside or outside the firm.

The odd part of it all is that implicit costs are almost always considered when accounting rules don't get in the way. Take a capital spending decision or a possible lease or buy move. In both instances the costs of money (both explicit and implicit) form an integral part of the final decision-making process.

3. *Capitalization versus expensing.* Here's still another accounting anomaly. Everyone takes for granted that the purchase of a piece of equipment constitutes a capital asset, and that it will show up later on the profit statement only as the machine is gradually written off. The theory here is that the firm is deriving benefits over a period of years, and hence should charge it off over this period of years.

But why not do the same thing with, say, a large advertising outlay? True, tax laws say we must regard this as a period cost instead of a capital asset. But isn't it equally true that an advertising campaign in period 1 will yield benefits to the company in period 2, period 3, and even beyond? If this is so — and it is — then why not capitalize rather than expense the advertising campaign?

The usual answer is that it is not easy to determine how much of the advertising outlay should be allocated to subsequent periods. And this is certainly true. But is it not also true that from a conceptual angle we are treating two outlays differently (one as a capital outlay and the other as an expense), even though we accrue benefits from both in subsequent periods? Maybe little can be done about this. But it should be recognized for what it is — a conceptual inconsistency stemming from the rigidity of our accounting approach to costs and profits.

4. *History versus potential.* Normally we gauge effectiveness by how profits have fared — particularly how they have fared in relation to past performance — taking into consideration, of course, the current business climate. Let profits show a good advance, and everybody is happy.

But does this really measure managerial effectiveness? Unfortunately, not always. Let's assume that this time you have the option of producing product A or product B. You opt for product

A and show a good-sized 30 percent profit advance. That's all the profit statement tells you. You know nothing about what would have happened if you had opted for product B. Perhaps your profit increase would then have been an even more impressive 50 percent.

The point is that earnings statements reveal little about profit optimization. And the reason is obvious.

Accounting records detail the history of a period—they tell nothing about what might have happened. The distinction is important, for in many cases it's only the latter information that can separate the men from the boys.

FIFTEEN

UNAVOIDABLE ERRORS

Ask five people to count the number of coins in a large glass urn. You're probably in for one of the bigger surprises of your life if you expect anything even approaching unanimity on the part of the counters. More than likely you'll wind up with five different answers. Indeed, it would be almost a miracle if more than two of the counters ended up with the same total.

The reason in this case is obvious. Counting several thousand coins is a tedious job—and the sheer drudgery and boredom of the operation are sooner or later going to cause most enumerators to err.

And therein lies a lesson for all users of the quantitative approach: Pure perfection is the impossible dream. No matter how careful one is—and no matter what steps are taken to eliminate distortion and deceit—a certain amount of error is unavoidable.

In the above case, an alternate might be sampling—counting the number of coins in one small portion of the urn—and inferring from this sample count the probable number of coins in the entire urn. But, as was pointed out in Chapter 7, sampling too invites the risk of error.

In short, there's no way out of the box. Nonpremeditated mistakes are a fact of life. Sometimes they can be traced to a simple cause, such as inaccurate or incomplete raw data, or even arithmetic errors involving their manipulation. At other times the time span or coverage can inject a note of uncertainty. Then, too, there are unexpected or one-shot factors that can throw an otherwise "scientific" forecast into a cocked hat. Finally, there are those elements that are unquantifiable—a situation that leads to subjective errors. Each of the above will be examined below.

INADEQUATE BASIC DATA

While there's a natural tendency to look for shady or unjustified manipulation of numbers and figures, often the fault lies in the original statistics themselves. We take for granted that the original data are reliable and seldom bother to ask pertinent questions about how they were collected, tabulated, and processed. But such a charitable attitude may not always be warranted.

Figures are particularly suspect where anticipation surveys are involved—surveys that ask businessmen or consumers how much they plan to spend over some specified future period. The biggest headache here is that people answer without really knowing or caring. It costs nothing to tell the questioner that you are going to buy that car you've had your eye on for months. But will you actually go through with the deal when it means shelling out thousands of dollars or committing yourself to large installment payments for upwards of three years?

The question is not an unimportant one, for history has shown there can be a big difference between what people say they are going to do and what they actually do. Nor should this criticism be limited to consumer studies. Businessmen, answering anticipation surveys about their future capital spending plans, are equally remiss. They might very well be planning to purchase a new machine tool, but at the last moment they may change their minds because of a shift in the business climate or simply because someone in the executive suite has had some second thoughts about the project.

In both the consumer and the business spheres, then, these data on future spending plans should be taken with a few grains of salt. Rarely are they 100 percent accurate. Indeed, errors of 5 percent or higher are more the rule than the exception.

All this in no way is meant to imply that such anticipation surveys are useless. Far from it. They usually do point in the right direction, even if the magnitude of the change is off a bit. Equally significant, they also serve as a gauge of consumer and business confidence—and are thus invaluable clues to general drift in these two crucial spending areas. On the other hand, they should be recognized for what they are: not precision barometers that can pinpoint tomorrow's trends, but rather nebulous numerical clues to be used in conjunction with other more precise types of statistical input.

Since most statistical information is based on surveys, it might also be well to point out that even historical data can be slightly askew. A consumer who is asked what he spent last month will not remember everything—or exactly what he paid. There could also be a tendency to forget one or two bad purchases. True, government figure collectors try to help out by providing forms and then checking out the results for consistency. But even so, it's hard to eliminate all such recall-type errors.

At still other times, surveys can go off the beam because of definitional or sampling problems. Unemployment offers a perfect example of these types of potential error. The common-sense definition of unemployment is simply someone out of work. Not so Uncle Sam's definition. The latter concept does not include people who are out of work but not looking for a job. Moreover, it bypasses those who because of frustration have given up the ghost and dropped out of the labor market. And it's likely that it underestimates many ghetto dwellers who either are so poor or so transient that they have never even appeared on a census list.

Still others criticize this widely monitored statistic on the grounds that it fails to capture seasonal shifts with any real degree of precision. The problem here is that figure collectors rely upon seasonal patterns observed over past years, and that, as such, they are unable to pick up any new pattern that might just be emerging at the time the survey is being taken. The fact remains that such seasonal factors as weather, holidays, production schedules, and school openings and closings are never entirely constant. And to the extent they're not, it means introducing a possible source of error.

How much error? It's hard to say on a month-by-month basis. But past revisions of seasonal adjustment factors suggest an average error of 0.1 percentage point in the unemployment rate.

Finally there's the sampling error to factor into any jobless figure. Since full enumeration is impossible (only about 100,000 are sampled each month), the possibility of chance errors can't be ignored. Even though the sample is representative and above statistical reproach, this means that a given level of unemployment could vary from month to month by close to 0.2 of a percentage point because of chance factors. In short, the unemployment rate must go up or down by 0.2 percentage points to be statistically significant.

Even a 0.2 percent change could conceivably be insignificant; for, according to government statisticians, chances are only nine out of ten that the true change in the jobless rate will vary within the 0.2 percent range. Thus in one out of ten cases, the 0.2 percent change could still be due to chance.

Sum up the 0.1 percent seasonal error and the 0.2 percent sampling error—and add in the controversy of who should be considered unemployed—and you see that there has to be a relatively large change in the reported figure to indicate any significant and meaningful change.

Are, then, such survey figures any good? The answer is still, yes. But their real value lies in their movement over a period of several months—when changes on the upside or downside have had a chance to accumulate. If unemployment has gone up 1 percent or down 1 percent over, say, a six-month period, you can be pretty well certain that the jobless situation has changed significantly.

On the other point before leaving the subject of inadequate basic data: Numbers arrived at via the residual approach are usually more suspect than other numbers. Thus in national income analysis, savings are what is left after private consumption is subtracted from disposable personal income. Since both of these are very large compared with the amount of savings, small errors in income and spending estimates can lead to large percentage errors in the estimate of savings.

ARITHMETIC ERRORS

Surprising as it may seem, many wrong numbers come up simply because the analyst or computer goofed. On the latter score, how many stories have you heard of people being billed for major appliances they never purchased—or of magazine subscribers who receive no copies for several months and then suddenly are inundated with three or four copies for each succeeding month? And what about the outrageously high electric or telephone bills some people receive on occasion? The favorite excuse for all the above: "Computer error."

The same might even be said of some government reports. More than once the monthly release of prices or production has been held up because the answer spewed out by the computer seemed unreasonable. Note, too, the number of revisions that are made after a few months have elapsed. Usually "Receipt of new data" is the excuse given. But talks with the men who are actually responsible suggest that a good portion of such changes is due to little more than sloppy computation, sparked in many cases by the rush to meet a deadline.

The problems of arithmetic or computer errors seem to increase with the complexity of the work being done. A few years back, for example, the government was working on a series of input-output tables which were supposed to trace all the interactions among hundreds of different industries. Admittedly, this was a formidable task, and when the government announced a one-month and then a two-month postponement, most people were willing to accept this as a reasonable underestimation of the time needed to complete the project.

But when continued postponements pushed the release more than a year beyond its original target date, this was something else again. The real truth: The computer had been improperly programmed—so much so that the tables being spewed out were both highly unbelievable and inconsistent in terms of their relation to other tables. An agonizing reappraisal had to be made before the difficulties were ironed out and a meaningful, consistent set of tables were developed.

In a sense, it may be unfair to place all the blame on the computer—or even the people running the computer program. Sometimes the errors are simple human ones—like the time when consumer sales were 10 percent too low because one whole reporting area had been inadvertently dropped out—or the time a major price index had to be revised upward after it was found that someone had (again inadvertently) fed last year's figures into the computer.

THE DEGREE
OF AGGREGATION

Estimate the sales of a particular company, and the odds of your projection being right on the nose are relatively small.

Estimate the sales of that firm's entire industry, and your chances are considerably better. Finally, make an estimate of the all-industry sales total, and your chances of hitting the bull's eye are still greater.

There's really nothing mysterious in all this, for the above is one of the many illustrations of the "law of large numbers." By considering many subgroups that make up a whole, any error in one subgroup is likely to be canceled out by a countervailing error in another subgroup.

That's why, for example, the government's consumer price index is more reliable than, say, its subindex for food or one of its metropolitan area subindexes. On the CPI front, for example, most statisticians agree that when this bellwether price yardstick rises 0.2 percent, we can be 95 percent confident that prices actually did rise. But when we get down to the subgroup level, changes of 0.3, 0.4, and 0.5 percent may be needed before we can be 95 percent sure that an advance occurred.

THE UNCERTAIN FUTURE

Just as the error factor is dependent on the level of aggregation, so is it also dependent on the time factor. You can generally predict what's going to happen tomorrow a lot better than you can foretell what will happen next week or next month. Other things being equal, the farther out you go, the less certain your projections. The reasons are rather obvious. For one, most of the clues for tomorrow's forecast are already in. Thus if it's tomorrow's production schedule that's to be projected, you have today's orders as a rather infallible guide. Contrast this with projecting next year's production schedules. You may have some idea what the order situation will be then (based on demand forecasts), but it's certainly nowhere near as definitive as when you were projecting tomorrow's schedule.

Secondly there's the assumption problem. We can be more certain that today's assumptions will hold up through tomorrow than we can be sure that they'll still be holding up a year from now. If our exports and imports are in balance, we can probably assume no devaluation over the next few months—and be reasonably sure we will be correct. But there's no telling what can happen on the international monetary front over the longer pull.

Who, for example, in 1970 could have forseen two consecutive United States devaluations—in December 1971 and February 1973? Indeed, as late as spring of 1971 most financial analysts were discounting any chances of such a move. (The rapidity with which many assumptions can change was discussed in greater detail in Chapter 8.)

There's still another factor working for decreasing accuracy over the longer pull. It stems from the fact that the basic relationships which determine a forecast tend to change over time. A few years ago a given level of copper water tube, for example, could reliably be predicted on the basis of a given level of construction activity. But within a few years the development of cheaper and equally reliable plastic tubing suddenly changed the entire relationship. Today copper plumbing is reserved for only the carriage trade. In short, long-term forecasts made only a few years ago suddenly became completely valueless because of a major shift in the copper tubing-construction relationship.

Government economists and corporations usually take cognizance of this phenomenon. They accept less accuracy over the longer pull—and because of this, they generally ask for aggregate rather than detailed forecasts over an extended time span. Thus a car manufacturer trying to estimate demand a decade from now would be interested only in the possible number of units he might sell. He would not want or expect any breakdown by size or model—primarily because such projections would be little more than guesswork.

THE PROBLEM OF
FORECAST FEEDBACK

This type of error, usually associated with projections, generally stems from not taking into account corrective measures that people may take to bring the future results more in line with basic targets. If, for example, profit projections are under desired levels, management may take new steps (not assumed in the forecast) to beef up the earnings level. Some statisticians refer to this as the "Heisenberg effect," after a famous physicist who stated that the process of measuring a phenomenon often changes the parameters of that phenomenon.

Care must be taken not to blame the analyst for variation stemming from this kind of feedback pressure. Indeed, if anything, an accurate forecast of low profits may have performed a needed service—alerting management to the fact that current policies were not consistent with the firm's basic goals. On the other hand, some of this error can be reduced if the analyst, when coming up with his first tentative forecast, checks with management or the government, or whoever else has commissioned the forecast, to see whether any change in the underlying assumptions may be necessary. If so, he can then change his forecast—and at the same time note that the more favorable forecast is premised on additional actions, which are then spelled out.

ONE-SHOT FACTORS

One of the nightmares faced by forecasters is the very real danger that some sudden, unforeseen event can turn all their careful forecasts and projections upside down. No amount of planning, no amount of foresight, can help in these matters — because, in a sense, they are "acts of God."

Indeed, many times it is just some unforeseen natural catastrophe that can upset the applecart. A drought in one part of the country can make for an unexpected food shortage and rising prices. A heat wave can send sales of beer and air conditioners skyrocketing. A hurricane can result in millions of dollars of unexpected insurance claims. Not only can't such catastrophes be foreseen, but unless they're recognized for what they are, serious errors in future calculations may result. A particularly hot summer which may have raised the sale of air conditioners above their normal rate of increase doesn't and shouldn't signify any shift in underlying trend. Similarly, a drought-inspired food price rise one year doesn't signal continuing rises in subsequent years. In both cases the results were atypical and must be recognized as such when future projections are being made.

The one-shot problem, however, is by no means limited to "acts of God." Thus a sudden change in government ground rules can equally make for sudden peaks or valleys.

A few years back there was talk to the effect that the government might soon suspend its investment tax program — a program that effectively reduced the price of many capital equipment items by about 7 percent. As soon as it became known that Uncle Sam was contemplating such a move, there was a rash of capital equipment orders — designed to get on the books before the rule was changed. Then when the rule actually was changed some months later, new business dropped off to a whisper. Neither the bulge nor the subsequent nosedive was normal, of course. And analysts who had predicted a steadier sales curve before the rumor got started could hardly be faulted for missing the boat.

As might be expected, the opposite occurred a few years later when the credit was restored. Business dropped off sharply when the rumors appeared, and then zoomed when the credit was actually put into effect.

Changes in income tax also have a way of distorting underlying patterns. Whenever a tax-rate decline is announced for a subsequent year, both individuals and corporations try to squeeze as many costs as possible into the earlier or higher-tax year. It's a surefire way of reducing tax bills over the two-year period. But it does have the unfortunate side effect of distorting corporate profits and government revenue estimates.

Even prices can be affected by legislative action. In 1973, a new federal law went into effect which made tampering with a car odometer (mileage indicator) a criminal offense. The predictable result: The number of low-mileage used cars dropped off precipitously. But with demand for these "cream puffs" remaining high, the market imbalance made for a sharp one-shot 6 percent rise in used car prices—all at a time new-car prices remained relatively steady.

This price rise in used cars, in turn, led to repercussions in the new-car market as well. Specifically, higher used vehicle quotas, combined with stable new-car prices, had the net effect of reducing the average "cost of trade" by about 3 percent— providing a powerful additional sales incentive for the purchase of a new car that year. Detroit, for example, estimated that the industry was able to sell 200,000 to 300,000 additional units that year because of the unexpected large rise in used car prices.

Hedging, too, can result in serious distortions of underlying long-term trends. Actually there are two precipitating factors. If a strike threatens in a major industry, there's a rush to buy to avoid getting caught short should the work stoppage actually materialize. At other times, there's a buying rush to avoid an announced price hike.

Thus the big 1973 bulge in car buying (which was partially attributable to higher used car prices) could also be partially traced to a widely anticipated $100 to $200 boost in the prices of 1974 models.

Here Detroit put the added "hedge" sales at near the 200,000-unit mark. If you add this to the 200,000 to 300,000 units attributable to higher used car prices, it means that anywhere from 400,000 to 500,000 units were sold in 1973 because of one-shot factors.

Businessmen at one time were also prone to price hedging. But this practice has diminished in recent years because of high inventory carrying charges. Since it costs anywhere from 2 percent to 3 percent a month to store additional supplies, a price rise has to be pretty stiff to justify extensive price hedging on the industrial purchasing front.

A few additional words are also in order about the second type of hedging—hedging against a possible supply dislocation. This is where businessmen enter with a vengeance. Up till recently buyers with clocklike predictability would begin to lay in a few extra months' supply of steel every third year—with the hedging spree coinciding with labor negotiations which took place in the industry during that year. This, of course, played hob with production and sales of steel. During the pre-hedge buildup, activity soared far beyond that which would have been justified on the basis of underlying consumption patterns. And subsequent to the usually peaceful settlement, activity dropped precipitously as industrial buyers rushed to pare their unneeded and expensive-to-carry inventory excess. To be sure, some have argued, "So what—didn't the hedge and pairing balance each other out over a year?" This is, of course, true. But only at tremendous expense—both to the buyer, who had to increase his carrying costs, and to the producer, who found the stop-and-go scheduling extremely inefficient from a production vantage point.

Another thing that bugged both producers and the labor unions alike: the rush to hedge invariably forced many users to look abroad for additional supplies. The problem here was that once these overseas producers had their foot in the door, they weren't about to get out.

So, for a variety of reasons, it became apparent that few were gaining from these periodic three-year steel hedging sprees. This being the situation, in 1973 both the steel producers and the United Steel Workers signed a no-strike pact—agreeing to settle any disagreement by compulsory arbitration. Hopefully this will spread to other industries—for hedging aids no one, adds to expenses, and reduces overall productivity. If producers and unions live up to their agreement, it could lead to the gradual demise of a one-shot factor that has been distorting business for more than four decades.

Fewer strikes would also assure a more orderly production process in other industries as well. The one-time steel workers did go out on a long strike (1959), it disrupted the whole economy, with auto companies and other steel users unable to get needed material. Result: Overall industrial production and GNP went into a tailspin—one that many now blame for the recession which subsequently took place.

Nor are one-shot interruptions necessarily limited to the domestic economy. We live in an increasingly related international world, and when one country sneezes, we all catch cold. Consider devaluation. Whenever rumors of such a move crop up, there's a rush into metals as a hedge. Gold, silver, platinum, and even some of the base metals move up in price far beyond the levels justified by supply and demand conditions—only to fall back when a new monetary equilibrium is established.

Such speculation also tends to build up during periods of

international tension. War scares result not only in higher precious-metal prices but also in the buildup of industrial supplies. And the buying sometimes spills over into the consumer area. The classic example of this came in 1950 to 1951, during the Korean war, when demand for autos and appliances took a sudden jump, only to fall back once the threat of global conflict began to diminish. Such consumer scare buying, however, has tended to drop off in recent years—either because consumers have become hardened to today's global conflicts or possibly because the real dangers aren't quite as intense as they were during the cold war years of the 1940s and 1950s.

There's still another international influence—this time economic—that warrants attention. It's the stop-and-go buying and selling approach of many government-controlled economies. Whether the Russians decide to sell or stop selling platinum is an imponderable that can play sudden havoc in precious-metal markets. The Chinese, too, have often tended to upset the applecart. Thus in early 1973, Chinese buying of large copper tonnages was one of the primary reasons for a major runup in this key market. And a few years earlier, an equally sudden Chinese withdrawal from the antimony market on the selling side sent prices up three and four times their normal level.

Political agreements or understandings can also affect data. The sharp drop in United States steel import tonnage during 1969 was directly attributable to a voluntary quota agreement on the part of foreign sellers—and not to any diminution in demand on the part of United States buyers. Similar pacts have made for sharp shifts in the importation of certain textile products.

DEALING WITH
THE UNPREDICTABLE

By its very definition, we will never be able to predict all the unpredictable and one-shot events outlined in the previous section. But there are things we can do to temper their effect — to reduce the possibilities of drawing any wrong inferences. Certainly if a wage contract is up next year, then in the absence of a no-strike pledge we can clearly expect major production and sales irregularities as the deadline day approaches.

But beyond this, about all we can do is call attention to the disrupting force when evaluating future market prospects. Thus there are some who would eliminate a certain period from their analysis if it were distorted by a strike. Or they might make an adjustment — they might push the data either up or down over the affected period to compensate for this one-shot event.

There are some other useful techniques for avoiding costly errors, too. If you're predicting sales of ice cream or soft drinks, it might be a good idea to think in terms of a temperature range during the key selling season and present your forecasts accordingly. Then there are some who use the *moving average* approach. Under this technique a given month's results are averaged with results of adjoining months. In this way any sudden or unexpected dip or rise is cushioned, permitting you again to get at the underlying trend.

THE NECESSARY SUBJECTIVE ELEMENT

Any quantitative report has to be seasoned with some qualitative judgment. Few analysts would deny this, and few would not admit to making some sort of subjective adjustment to numbers and figures before they are released to the public. A problem arises as to just what kind of subjective adjustment. Certainly no two analysts will agree to the last decimal point.

This, then, can inject still another unavoidable source of error — because, by definition, there's no way of knowing how correct or incorrect any such subjective judgment may be.

Hindsight often reveals that the error was nothing more than a case of poor judgment. At other times it may be that the analyst can only guess—or that clues as to the extent and direction of any needed refinement are simply not available. At still other times, a definite bias may exist—a bias which, incidentally, may be unrecognized by the people professing to give you the pure, unadulterated truth.

Such biases can sometimes lead to highly questionable results. We've all met people who are so obsessed with what they want to prove that they are unaware of the obvious blinders they're wearing. They will use all the data at their disposal to establish their point of view, but be totally unaware of other statistics which might raise some questions about their underlying thesis.

Read a right-wing newspaper, for example, and you will have more than enough statistical ammunition to back up any conservative point of view. Similarly, a left-wing newspaper will provide equally impressive documentation of why the liberal path is the only true and correct one. The only real solution for the reader, of course, is to read the evidence in both papers, weigh the pros and cons, and come up with a conclusion based on this broader range of quantitative intelligence.

A similar situation might apply at the labor bargaining table, with both the unions and management each seemingly presenting a set of impeccable statistics to back up their point of view. It's only when you see both sets that the true, unbiased picture begins to emerge.

Bias can also crop up when the original data are being collected. This was pointed up earlier in the chapter when consumer responses to surveys were being examined. But the latter type of bias can be at least partially dealt with by correcting responses in line with historical experience—although admittedly these corrections, which in themselves are partly subjective, have up to now been far from perfect.

Still another type of bias is worthy of note—the type that leads people to err on the conservative side. A market forecaster will generally tend to underproject rather than overproject. And with good reason. If sales top his levels, then everybody's happy and his mistake gets lost in the general feeling of euphoria.

On the other hand, if actual sales fall below projected levels, everybody is up in arms. Why did the company fall short of target? Was it the salesmen's fault, the market researcher's fault—or somebody else's fault? A scapegoat must be found—and it might just as well be the person who built up our expectations, only to dash them later on.

No wonder, then, that a forecaster is a little less than totally honest. It's safer to err on the conservative side. Why crawl out on a limb when you don't have to? It's little more than our instinct of self-preservation.

But we can do something about this kind of subjective error. Let's not make it more threatening or less rewarding to overestimate than to underestimate. Let's change the rules of the game—let's penalize errors on the plus side equally as hard as those on the minus side. In short, the answer may be to apply quantitative rather than qualitative yardsticks to forecasting performance.

But any such reform won't come easy. There's an inherent resistance to change in all of us. Thus if the time-tested relationship between sales and income suggests x volume of sales, then it will be x sales that will be shown in the projection—despite the fact that we suspect that the relationship may be changing. This same resistance to changing a relationship recently cropped up in capital spending projections. The old spending relationship was closely tied to capacity utilization—with only minor weight being given to such other determinants as labor costs, obsolescence, and competitive pressures. But these latter factors have recently become much more important— so much so that analysts using the old relationship have been consistently underestimating outlays. The usual problem: Nobody was willing to stick his neck out.

Similar problems are behind some of today's shortages. Thus both the telephone companies and the electric utilities have tended to underestimate demand—with resultant shortages and service interruptions. Airport planners have been guilty of similar underestimates. And how many times have you seen a new highway that has just been completed found to be inadequate to handle the volume of traffic it was planned for?

To be sure, all these miscalculations can be explained away in terms of upward shifts in growth curves. But this is hardly any excuse. If you're projecting future trends, all the factors that might possibly influence future growth patterns should have been quantified into your forecast. The fact that they haven't again suggests an inherent conservative bias on the part of the forecasters.

PSYCHING THE FUTURE

Before leaving the realm of subjectivity, a few words are also in order on our inability to quantify everything. Chapter 5 devoted considerable space to this overall problem. But here the accent is on the fact that while we may be able to quantify many aspects of a situation, we as yet cannot foretell that part of a number that depends on the reaction of individuals.

Much as we like to think otherwise, we are not completely rational beings. One year an income may evoke a given spending pattern, while another year with the same income may result in something considerably different. Despite advertising and other manipulative measures, we have yet to develop the fully predictable consumer.

The same might be said of the consumer's counterpart, the

businessman. Company A may develop a detailed set of plans based on the most reasonable reactions of his competitors. Yet in an uncomfortably large number of cases his competitor will cross him up.

Businessmen have tried to overcome this hazard with such statistically sophisticated approaches as *market simulation*—where all the probable reactions of the competition are carefully weighted. But the results have not always been successful. If there's any rule that can be given on this score, it is never to underestimate the competition.

SIXTEEN

YARDSTICKS FOR ACCURACY

Beauty is in the eye of the beholder. So, to some extent, is accuracy. While everybody agrees that 100 percent accuracy is virtually impossible, the question arises, At what point do we draw the line between a "good" and a "bad" number or forecast?

A prominent government official tells us that prices will rise by only 3 percent in a certain year. The actual increase turns out to be $3\frac{1}{2}$ percent. Was his forecast reasonable—or did he miss the boat completely?

Similarly, an auto maker tells us that using his car (instead of his competitor's) can reduce annual fuel consumption costs by 10 percent. After tests are run, the actual savings turns out to be 6 percent. Was the auto maker reasonably accurate—or was his statement just another advertising pitch?

Unfortunately, even the term "accuracy" is subject to many different interpretations. To some, accuracy connotes absolute precision as far as the numbers are concerned. Yet to others, calling the tune on the direction of change rather than the precise numerical change is the basic criterion of accuracy. If the weather bureau calls for a sharp rise in temperature—say, from 75° F one day to 90° the next—and the actual rise is only to 85°, the majority of the people would call the forecast reasonably accurate. They would only deem the forecast off-base if temperatures remained unchanged or actually fell.

In short, in many cases direction as well as numerical change must be considered in assessing the validity of any number. Take another hypothetical example—this one involving two price projections—one calling for a 3 percent rise over a stated period of time and the other calling for a 1 percent decline. As it turns out, the actual result is a 1 percent rise. Which is the more accurate of the two forecasts?

From a purely numerical perspective, they are about equal— one is 2 percentage points too high while the other is 2 percentage points too low. But in terms of usefulness, the first one is almost always deemed to be the better one—for only this one called the direction of change correctly.

This "direction" element is not to be underestimated. If you're running a business, for example, pinpointing the direction of change can be of invaluable assistance in setting inventory policy. One large retail chain tells of the time it opted for the seemingly "way-out" forecast that called for business improvement. As it turned out, the forecast was correct, and only this chain had enough merchandise on hand to meet the heavier consumer purchases which materialized. Result: Its sales rose while competitors, caught with their inventory pants down, actually suffered a continuing decline in sales.

There's still one other general aspect of accuracy that deserves note. One forecast may have been correct on pure guesswork—with the intellectual underpinning that led to the final numbers deemed to be questionable at best. Compare the above with a similarly accurate forecast where the basic relationships among the variables in question were established with impeccable statistical precision. Clearly, the second forecast is more valid. And when these two forecasters come out with their next projection, we would certainly be justified in putting more faith in the second approach.

All this sounds obvious. Yet how often do forecasters make a name for themselves by "shooting from the hip." Every once in a while they score a bull's-eye. And on the basis of this one lucky shot, they manage to sell their forecasting services to some of the nation's leading corporations, who then find out a year or so later that their thousands of dollars in "consulting service" costs are bringing them little more than statistical static.

The point here is that validity should be an integral part of any forecast evaluation. It's worth the time and effort to look into how the forecast was built up. Without a solid, logical foundation, no forecast is worth the paper it's printed on.

To sum up, then, "accuracy" is a relative term. Looking at the final numbers is never really enough. What's almost always needed is the establishment of a set of objective criteria—ones that can separate the useful forecasts from the spurious ones, the reliable ones from the purely lucky ones. Season all the above with a pinch of subjective appraisal, and you're well on your way toward meaningful yardsticks of statistical accuracy.

A more detailed rundown on some of the good and not-so-good appraisal techniques follows:

ACTUAL VERSUS PROJECTED

This is simply the yardstick which measures the numerical difference. As was pointed out above, one of its basic shortcomings is that it fails to measure sensitivity to directional change. There are actually two forms of this basically faulty gauge of accuracy. The first involves calculation of the absolute difference; the second, calculation of the relative difference.

While both are defective, the relative difference approach at least attempts to put the observed numerical gap into somewhat better perspective. Thus if General Motors is $1 million under its yearly sales estimate, it's a lot different from the same $1 million gap observed for the XYZ Widget Co.—for the $1 million GM shortfall may represent only a fraction of 1 percent of its total sales volume, while the XYZ Widget shortfall could conceivably represent 10 percent or 20 percent of its yearly sales volume.

Obviously the relative or percentage approach is the only meaningful one in this case, because the size of the firms being compared is so different. Yet, obvious as all this seems, there are people who still cling to the "absolute gap" approach— either because of ignorance or because they hope to make a stronger case for their side by doing so.

But the relative yardstick approach—aside from its insensitivity to directional change—can be faulted on still another ground. While it does give some degree of standardization (in that it can be used to compare series of different magnitudes), it leaves one important question unanswered: The error may be 2 percent, 4 percent, or 6 percent; but how does one decide whether any of these percentages are good or bad? The following approach can perhaps best provide the answer.

THE PROBABLE ERROR

This is a yardstick, based on the laws of probability, that (1) measures the expected variation about a projected number and (2) gives the forecaster the probability of getting such variation. Statisticians refer to this as the "standard error" approach. It can be shown, for example, that 1 standard error on either side of a projection should account for about two-thirds of all chance variation; 2 standard errors on either side of the projection should account for 95 percent of the chance error; and 3 standard errors on either side should account for virtually all possible chance errors.

This measure makes it obvious that an actual reading of, say, 5 standard errors could not possibly be due to chance. If such a large standard error should appear, it is clear that the estimating technique is poor—that the error is much too big to be accounted for by normal chance factors.

The standard error technique also eliminates one serious objection that can be levied at all the yardstick approaches previously discussed: It is the only one that takes into account the different inherent variabilities of the data being monitored. Thus the standard error in estimating sales for a highly volatile industry (such as autos) would very likely be a lot larger than one in a relatively stable industry such as food. Indeed, the increase in the auto industry's standard error would be directly proportional to this industry's greater inherent variability. Put another way, in the case of the more volatile auto forecast, the analyst could automatically set wider limits over which the actual results could be expected to vary from projected levels on the basis of past experience.

The calculation of this standard error is beyond the scope of this book—although it can probably be understood by anyone who has had a beginning course in college statistics. But more important than the calculation (to the layman, at least) is the knowledge that such a measure exists. It suggests that when forecasts are given, they should always be accompanied by the standard error—to give the recipient a hint as to their reliability and probable range of error.

Lest there still be some doubt as to the necessity of providing a gauge of the chance factor, consider the following exaggerated example involving two meteorologists. One happens to be located in the desert, where it almost never rains, and the other is located in an area of variable weather patterns.

Clearly, the desert forecaster, if only by predicting "fair and warm" every single day of the year, is going to rack up a better record than his counterpart who has to deal with major uncertainties every day. In any event, it would be unfair to compare meteorological performance without considering the sharply different "variation" factor faced by the two forecasters.

PUTTING THE ACCENT ON CHANGE

So far most of the emphasis has been on deviation—either relative or absolute—from observed results. But there is still another way of looking at error: a comparison of projected change with actual change. An example can perhaps make the distinction between the two approaches clearer. If you predict a rise from 100 to 102, and the actual result is 101, then your absolute error is 1—and you are about 1 percent off target (the actual result of 101 compared with the projected result of 102). But put the accent on change, and the results suggest a much larger error. Specifically, compare the projected change (2) with the actual change (1), and you end up with a "change" error of 100 percent—that is, your projected change was double, or 100 percent more than, the actual change.

Indeed, when errors are expressed in terms of anticipated change versus actual change, the deviations are always much larger, thereby providing a more sensitive yardstick for distinguishing between good, acceptable, and poor forecasts. Many analysts prefer the "change" concept for another reason, too. They point out that the aim of any forecast is to predict change;

therefore any evaluation of such predicted change should also be in terms of change.

STANDARD REALIZATION YARDSTICK

This recently introduced measure gives the advantage of evaluating accuracy in terms of both (1) inherent variability of the numbers being analyzed and (2) the ability to predict a turning point.

Again the arithmetic is much too complex to discuss here. But the payoff is certainly clear-cut. As one top researcher, Robert L. McLaughlin, puts it, "The standardized realization (SR) percent standardizes most of the things we want to know within the concept of a single number. It describes accuracy in terms of turning points, direction of change, amount of change and implicitly hints at the extent of fluctuation." On the latter score, it doesn't matter whether one is dealing with a series that fluctuates a lot or a little—or whether it be in dollars, units, tons, pounds, gross, percents, etc.

OTHER ASPECTS OF ACCURACY

1. *Beware of the super-hedger*—the one who invariably claims credit when one of his many alternatives turns out to be correct. This, of course, does no one any good—unless, of course, the recipient has the subjective intuition to know which version to use. A story making the rounds just after World War II illustrates the worthlessness of this multiple-forecast approach.

A senior Nazi official was asked how good German intelligence was on the Normandy landings. "Perfect," he replied. "We knew the day, the hour and the beaches. Unfortunately, the right estimate could not be distinguished from the other 40 that were wrong."

2. *Make use of "naïve" models.* Where possible, compare

projected results with those obtainable from simpler models or forecasting techniques. If you can't do better than such naïve approaches, then it's a good idea to think about (1) upgrading your techniques or (2) reverting to simpler ones which seem to yield almost as good results.

3. *Don't forget common sense.* Essentially this involves asking the question: Does the conclusion seem reasonable in view of what is already known? Sometimes, for example, straight extrapolation of past trends can lead to ludicrous results.

For example, if one had taken the percentage increase in the number of TV sets between 1947 and 1952 (10,000 percent) and extrapolated this percentage another five years out, one might have concluded that there would be several billion TV sets in the United States by 1957. This would have meant 10 or more TV sets for every man, woman, and child—clearly a ridiculous conclusion. There are many less obvious types of this kind of blind extrapolation. The fact is that general knowledge of an industry—industry savvy, if you will—is a must for intelligent statistical knowledge and evaluation.

4. *Mix in some elements of subjectivity, too.* A too rigorous application of quantitative yardsticks—no matter how carefully formulated—can also lead to mistakes and misinterpretations. In many cases, for example, the conditions under which the forecast is being made should play a relevant role. Thus a family that is just about balancing its income and expenses will, other things being equal, require greater accuracy in forecasting its outlays. A slight variation in such a case could make a difference between solvency or going into debt.

The same theme carries over to the business sphere. If millions of dollars are riding on a number, then it had better be more accurate than one involving only a minute fraction of a corporation's business.

In some cases, even the corporate response to an error can play a role in determining the degree of accuracy required. If management reaction time to error is short, then obviously you might be able to squeak through with a somewhat less accurate figure than you might normally need. On the other hand, a slow response (because of red tape or the large number of management layers involved in any final decision) would require greater-than-average stress on accuracy.

In short, projections should stand or fall on the imperatives of the situation as well as numerical yardsticks. One thought to always keep in mind: If the numbers are accurate enough to meet your needs, then it can be deemed a useful forecast.

IMPROVING ACCURACY

No matter how good a number or projection turns out to be, it can usually be made even better. Areas worthy of exploring in pursuit of this ambitious goal might include:

1. *Installing systematic checkups and postmortems.* Its always a good idea to see if things are proceeding according to schedule or projection. One thing that emerges from such audits is the fact that revisions are sometimes necessary, and when effected they should be regarded as another service of forecasting rather than as a black mark for having erred in the first place; for in the long run, the more up to date a forecast is, the more useful it will be. There certainly should be no stigma attached if the change is due to unforeseen developments.

Postmortems have proved extremely useful, too. If nothing else, they stimulate thinking. In a great many cases such after-the-fact reviews have helped point up the reasons for any observed errors. And this is generally regarded as "must" information for any subsequent effort to improve forecasting intelligence.

2. *Upgrading data and techniques.* Much of the initiative to improve upon the raw numbers must in the end come from users rather than from the purveyors of the original data. For these users are the people who have most to lose by relying upon

inaccurate or meaningless figures—and they have the most to gain by improving upon them. Three means of such upgrading of statistical intelligence suggest themselves: (1) the finding of more accurate sources, (2) the "adjustment" of data already available, and (3) the possibility of obtaining the needed information a bit sooner.

The first of the above is self-evident—with the main caveat here being to avoid the trap of thinking a figure is accurate because it's in print. The second type of upgrading essentially involves correcting any raw data to take account of any seasonal, price, or other factor that might distort the underlying trend. The final suggestion—finding more up-to-date data—can often play a crucial role in improving forecast accuracy. Thus one additional month of income statistics may well reduce your sales forecasting error by 1 percent or 2 percent. Similarly, receipt of company price performance even a few days earlier than normal could conceivably spell the difference between success and failure of near-term marketing strategies.

The advent of the computer also has opened up vast new potentials. No longer need you be wedded to the old-fashioned, simple techniques. Some of the most subtle approaches, such as model building, are now within the reach of almost anyone with a little money and time.

3. *Deciding where statistical improvements can yield the biggest payoff.* The growing complexity of the world we live in always seems to be dictating an upgrading of our numerical intelligence. But given the demands of money, time, and other resources, it is highly unrealistic to aim for improvements on all fronts at the same time. Where, then, to concentrate?

The key question always to be explored: Is the effort worthwhile? The payoff must be impressive enough to justify the time and cost of installing more sophisticated data processing equipment and techniques.

Then, too, it's usually a good idea to concentrate on existing weak areas. If you know that many of the past projections have constantly missed the boat in a given area, why not put your initial upgrading efforts in this sector? Other questions that should be asked in any statistical upgrading effort: (1) Is the area you want to improve upon amenable to the quantitative approach; and (2) Is it likely that management, government, or whoever holds the purse strings can be sold on your improvement strategy? In the case of company forecasts, for example, a skeptical management can scuttle the most impressive of new forecasting and projecting techniques.

4. *Making the numbers more useful and realistic.* The barrage of figures that we are all subject to are virtually useless unless they can be put to some positive use. Being told what the average price increase is likely to be over the next 12 months—or the GNP increase—or even the price of meat—is a waste of time unless plans are made to utilize the information in your overall living strategy.

Businessmen are notorious demanders of statistical intelligence which is seldom utilized. In some cases the figures are ordered just for "show." Plaster a wall with charts, and it looks pretty impressive when the boss walks in. Similarly, quote some statistic at a board meeting (even a partially irrelevant one), and someone is likely to be impressed.

Realism is an equally pressing problem. Sometimes a number or projection can be prepared in good faith, but it just doesn't mesh with existing conditions. Thus a nation might promise its citizens one million new cars for a given year—but then will have to tone down this estimate simply because it does not have the production facilities to meet this projected level.

In a similar vein, a forecaster must consider such other factors as the financial position of people and companies. On the personal side, willingness to buy must not be confused with the ability to buy. And on the corporate side, a firm might be able to sell more, but it may be able to do so only by overextending itself financially—something that could well offset the advantage of higher sales.

Then, too, a well-meaning legislator may convince himself of the necessity of spending $x billion to improve the quality of the environment. His figuring is impeccable. The only trouble is that his fellow lawmakers may have a completely different opinion on where available funds should be spent. Ergo, his projection isn't worth the paper it's printed on. In short, beware of equating a "what's needed" number with the more pragmatic "what's likely to be funded" number.